Advances in Biomedical Image Processing and Analysis

Advances in Biomedical Image Processing and Analysis

Editors

Michalis Vrigkas
Christophoros Nikou
Ioannis A. Kakadiaris

MDPI

Basel • Beijing • Wuhan • Barcelona • Belgrade • Novi Sad • Cluj • Manchester

Editors
Michalis Vrigkas
University of Western
Macedonia
Kastoria
Greece

Christophoros Nikou
University of Ioannina
Ioannina
Greece

Ioannis A. Kakadiaris
University of Houston
Houston
USA

Editorial Office
MDPI
St. Alban-Anlage 66
4052 Basel, Switzerland

This is a reprint of articles from the Special Issue published online in the open access journal *Applied Sciences* (ISSN 2076-3417) (available at: https://www.mdpi.com/journal/applsci/special_issues/Biomedical_Image_Processing_Analysis).

For citation purposes, cite each article independently as indicated on the article page online and as indicated below:

Lastname, A.A.; Lastname, B.B. Article Title. *Journal Name* **Year**, *Volume Number*, Page Range.

ISBN 978-3-0365-9638-9 (Hbk)
ISBN 978-3-0365-9639-6 (PDF)
doi.org/10.3390/books978-3-0365-9639-6

Contents

About the Editors

Michalis Vrigkas

Dr. Michalis Vrigkas is an Assistant Professor at the University of Western Macedonia in the Department of Communication and Digital Media, Kastoria, Greece. He received his Ph.D. from the Department of Computer Science and Engineering, University of Ioannina, Greece, and his M.Sc. and B.Sc. in Computer Science from the same institution. For the academic year of 2018–2019, he was an adjunct lecturer at the Department of Computer Science and Engineering, University of Ioannina. In the past, he spent two years (2016-2018) at the University of Houston, TX, USA, where he worked as a postdoctoral fellow at the Computational Biomedicine Lab and at the Texas Institute for Measurement, Evaluation, and Statistics. He has participated in several EU- and national-funded ICT research projects, while in recognition of the contribution made to the quality of several international journals, he has twice received the award of Outstanding Reviewer. He was the recipient of the Best Paper Award at the IAPR/IEEE International Conference on Biometrics (ICB) in June 2016. In addition, NVIDIA Corporation helped to support his research by donating a Titan Xp graphics card. His research covers a wide range of topics such as augmented and virtual reality, computer vision, image and video processing, image analysis, and machine learning with applications to medical image analysis and biometrics as well.

Christophoros Nikou

Dr. Christophoros Nikou received the Diploma in electrical engineering from the Aristotle University of Thessaloniki, Greece, in 1994 and the DEA and Ph.D. degrees in image processing and computer vision from Louis Pasteur University, Strasbourg, France, in 1995 and 1999, respectively. He was a Senior Researcher with the Department of Informatics, Aristotle University of Thessaloniki in 2001. From 2002 to 2004, he was a Research Engineer and Project Manager with Compucon S.A., Thessaloniki, Greece. He was Lecturer (2004-2009), Assistant Professor (2009-2013), and Associate Professor (2013-2018) with the Department of Computer Science and Engineering, University of Ioannina, Ioannina, Greece, where he has been a Professor, since 2018. During the academic year 2015-2016 he has been a Visiting Associate Professor at the Department of Computer Science, University of Houston, USA. Prof. Nikou was the General Chair of IEEE International Conference on Image Processing 2018 (ICIP'18). Since 2019 he is a member of the Conferences Board of the Signal Processing Society of IEEE. He is also an Associate Editor for IEEE Transactions on Image Processing. His research interests mainly include computer vision, pattern recognition, image processing and analysis and their application to medical imaging. He is a member of EURASIP and an IEEE Senior Member.

Ioannis A. Kakadiaris

Professor Ioannis A. Kakadiaris, Ph.D., is a Hugh Roy and Lillie Cranz Cullen University Professor of Computer Science, Electrical and Computer Engineering, and Biomedical Engineering at the University of Houston (UH), Houston, TX, USA. He joined UH in August 1997 after a postdoctoral fellowship at the University of Pennsylvania. He earned his B.Sc. in Physics at the University of Athens in Greece, his M.Sc. in Computer Science from Northeastern University, and his Ph.D. at the University of Pennsylvania. He is also the founder and director of the Computational Biomedicine Lab. His research interests include biometrics, computer vision, and pattern recognition, biomedical image analysis, and cardiovascular informatics. Dr. Kakadiaris is the co-founder of the Pumps and Pipes initiative which examines cross-domain innovation in the domains of medicine, the oil industry, and aerospace. He has authored more than 300 publications in international journals, and he holds twelve US patents. He is an international expert in biometrics, data/video analytics, and biomedical computing. His team has made contributions in the areas of 3D face (and ear) recognition, 3D-aided 2D face recognition, and profile-based face recognition. In addition to twice winning the UH Computer Science Research Excellence Award, Ioannis has been recognized for his work with several distinguished honors, including the NSF Early Career Development Award, the Schlumberger Technical Foundation Award, the UH Enron Teaching Excellence Award, and the James Muller Vulnerable Plaque Young Investigator Prize. His research has been featured on Discovery Channel, National Public Radio, KPRC NBC News, KTRH ABC News, and KHOU CBS News.

Preface

Biomedical image analysis plays a vital role in diagnosing numerous pathologies ranging from infectious diseases to cancer. Advanced methodologies for signal and/or image processing and analysis and biomedical analytics may be a powerful tool for classifying medical data, reasoning individualized health trends, and finding evolutionary trajectories between normal and non-normal cases in many medical applications. This Special Issue includes some of the latest research regarding biomedical image analysis and computer-aided diagnosis, reports novel imaging methods with biomedical applications, and explores the development of new algorithms for biomedical image processing and analysis. The main goal of this Special Issue is the dissemination of scientific results and innovative ideas among the scientific community and to bring different facets of health monitoring together.

Michalis Vrigkas, Christophoros Nikou, and Ioannis A. Kakadiaris
Editors

applied sciences

MDPI

Article

Optimizing Point Source Tracking in Awake Rat PET Imaging: A Comprehensive Study of Motion Detection and Best Correction Conditions

Fernando Arias-Valcayo [1,*], Pablo Galve [1,2,3], Jose Manuel Udías [1,2], Juan José Vaquero [4,5], Manuel Desco [4,5,6,7] and Joaquín L. Herraiz [1,2]

1 Grupo de Física Nuclear, Departamento de Estructura de la Materia, Física Térmica y Electrónica & Instituto de Física de Partículas y del Cosmos, Universidad Complutense de Madrid, 28040 Madrid, Spain; pgalve@ucm.es (P.G.); jose@nuc2.fis.ucm.es (J.M.U.); jlopezhe@ucm.es (J.L.H.)
2 Instituto de Investigación Del Hospital Clínico San Carlos (IdISSC), Ciudad Universitaria, 28040 Madrid, Spain
3 Paris Cardiovascular Research Center, Inserm UMR970, Université de Paris, 75015 Paris, France
4 Departamento de Bioingeniería, Universidad Carlos III de Madrid, 28911 Leganés, Spain; jjvaquer@ing.uc3m.es (J.J.V.); desco@hggm.es (M.D.)
5 Instituto de Investigación Sanitaria del Hospital Gregorio Marañón, Unidad de Medicina y Cirugía Experimental, 28009 Madrid, Spain
6 Centro Nacional de Investigaciones Cardiovasculares Carlos III (CNIC), 28029 Madrid, Spain
7 CIBER de Salud Mental Instituto de Salud Carlos III, 28029 Madrid, Spain
* Correspondence: farias02@ucm.es

Abstract: Preclinical PET animal studies require immobilization of the animal, typically accomplished through the administration of anesthesia, which may affect the radiotracer biodistribution. The use of ^{18}F point sources attached to the rat head is one of the most promising methods for motion compensation in awake rat PET studies. However, the presence of radioactive markers may degrade image quality. In this study, we aimed to investigate the most favorable conditions for preclinical PET studies using awake rats with attached point sources. Firstly, we investigate the optimal activity conditions for the markers and rat-injected tracer using Monte Carlo simulations to determine the parameters of maximum detectability without compromising image quality. Additionally, we scrutinize the impact of delayed window correction for random events on marker detectability and overall image quality within these studies. Secondly, we present a method designed to mitigate the influence of rapid rat movements, which resulted in a medium loss of events of around 30%, primarily observed during the initial phase of the data acquisition. We validated our study with PET acquisitions from an awake rat within the acceptable conditions of activity and motion compensation parameters. This acquisition revealed an 8% reduction in resolution compared to a sedated animal, along with a 6% decrease in signal-to-noise ratio (SNR). These outcomes affirm the viability of our method for conducting awake preclinical brain studies.

Keywords: positron emission tomography; awake PET; Monte Carlo; delayed window; random coincidences; anaesthesia; motion correction

Citation: Arias-Valcayo, F.; Galve, P.; Manuel Udías, J.; Vaquero, J.J.; Desco, M.; Herraiz, J.L. Optimizing Point Source Tracking in Awake Rat PET Imaging: A Comprehensive Study of Motion Detection and Best Correction Conditions. *Appl. Sci.* **2023**, *13*, 12329. https://doi.org/10.3390/app132212329

Academic Editors: Ioannis A. Kakadiaris, Michalis Vrigkas and Christophoros Nikou

Received: 9 October 2023
Revised: 10 November 2023
Accepted: 11 November 2023
Published: 15 November 2023

1. Introduction

Positron emission tomography (PET) is a powerful tool for imaging biological processes in vivo. PET scans can provide valuable information about molecular mechanisms of disease, drug safety and efficacy, and the response to treatments. In preclinical PET, animal models such as non-human primates and rodents are commonly used to develop and validate novel radiotracers and investigate disease mechanisms. However, the use of anesthesia during preclinical PET scans can have pharmacological effects that may affect physiological parameters, potentially leading to confounding results that limit the translation of preclinical results to the clinic [1–4].

To overcome this limitation, the focus on conducting studies with awake animals has gained importance in recent years. Several approaches have been taken in this field, including the use of restraining devices [5–9], scanners attached to the animals [10,11], and motion tracking and correction techniques. Restraining animals during PET scans can limit the animal's movement but may result in immobilization stress, leading to altered uptake of radiotracers [5,6]. Scanners attached to animals, such as the RatCAP [12,13], offer an alternative approach but may also induce stress in the animal. Motion tracking and correction techniques are currently the most-studied approach, as they allow free animal motion and ensure that the animal is not stressed during the scan.

Within the field of motion tracking and correction, several methods have been proposed. Optical markers [14–17], natural head features [18], point clouds [19], and point sources [20–22] are some of the most widely studied techniques. Optical markers require the rat's head to be facing the tracking camera, and there may be some limitations when the marker is occluded or the bore of the scanner is small. Using natural head features eliminates the need for attaching markers, but to obtain enough distinctive features, it may be necessary to paint a black pattern on the animal's head. Point clouds use a combination of stereo vision and structured light projection to represent the 3D surface of the animal head as point clouds, which can then be used to determine its 3D pose. Finally, point sources attached to the rat head have been widely studied by Miranda et al. [3,20]. This method uses the spatial location of the point sources in the PET data to calculate the head's pose [3]. This approach has shown promising results, making viable the acquisition of awake animal data without requiring any external devices.

Given the potential of using point source markers for motion estimation, our study aimed to optimize the parameters of point source markers strategy, a novel approach in motion tracking and correction. Our primary focus is to evaluate the performance of this motion-correction strategy, both with numerical phantoms as well as with several acquisitions with awake rats, including a reference acquisition of an anesthetized rat for comparison. The purpose of this research is to assess the performance of that method, particularly in situations where motion correction may be challenging. We also investigate when the activity of the point source markers may affect image quality. With these goals in mind, we have conducted a comprehensive analysis to enhance the effectiveness of the point source method and provide valuable insights for its practical applications.

2. Materials and Methods

2.1. Scanner

We have tested our mehtods in a 6R-SuperArgus [23]. The scanner is made up of two layers of 13 × 13 crystal arrays, each with a crystal pitch of 1.55 mm. The front layer consists of 7 mm-long lutetium–yttrium orthosilicate (LYSO) crystals, while the back layer has cerium-doped 8 mm-long gadolinium orthosilicate (GSO) crystals. The scanner has a total of 6 rings of 24 detectors each, with a radial field of view (FOV) of 17 cm and an axial FOV of 15 cm. Additionally, the scanner acquires data in a single list-mode, with information on the energy, time, and position of each event recorded.

2.2. Point Source Tracking and Motion Compensation

The overall workflow of the reconstruction process is divided into five steps (see Figure 1):

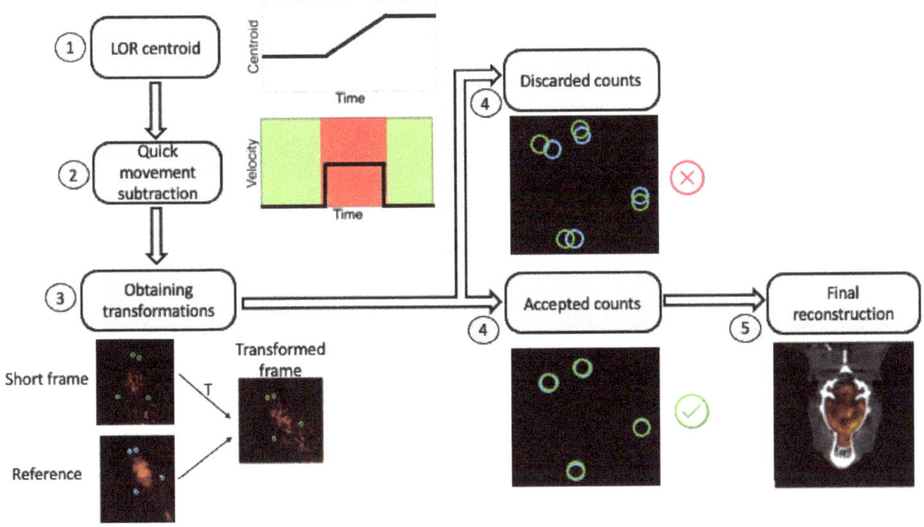

Figure 1. Workflow of the reconstruction process with motion tracking and compensation.

The reconstruction process is divided into five distinct steps, which are described below:

1. **LOR Centroid:** To address motion-related issues, we track the rat's movement during the scan. The centroid position of all LORs is calculated every 50 milliseconds, representing the movement center;
2. **Quick-Movement Subtraction**: Rapid movement is identified using the v_{max} parameter, derived from periods with minimal centroid variation. Such periods are indicative of minimal rat motion. Removing these high-movement intervals helps reduce motion artifacts, vital for small animal studies;
3. **Obtaining transformations**: We use a reference image from the most stable part of the scan. The acquisition is divided into 12.5 ms frames, reconstructed with low iterations while considering rapid movement removal. Rigid transformation matrices are derived by comparing point source locations with the reference;
4. **Non-Precise transformations subtraction**: To assess the quality of our transformations, we calculate a discrepancy measure, χ^2_{fr}, for each frame:

$$\chi^2_{fr} = \frac{\sum_{s}^{N}(p_s^{ref} - T(p_s^{fr}))^2}{N} \tag{1}$$

In this equation, N represents the total number of point sources, p_s^{ref} is the position of source s in the reference image, and $T(p^{fr}s)$ is the position of source s in a specific frame fr after applying the transformation T. Frames with χ^2 values below a set limit ($\chi^2 max$) are retained, as rigid transformations may not fully account for the rat's flexible skin, ensuring more accurate image reconstruction;

5. **Final reconstruction**: With the transformation parameters obtained for all frames, we proceed with the reconstruction process. Each event within a frame is transformed based on its corresponding transformation, adjusting scanner positions. As the scanner position changes during reconstruction, we need to adapt the standard Expec-

tation Maximization Maximum Likelihood (EMML) algorithm to ensure accurate reconstruction. We modify sensitivity corrections a_{ij} as follows:

$$a_{ij} = \frac{1}{T_{acq}} \int_0^{T_{acq}} T(t) a_{i'j} dt \tag{2}$$

where T_{acq} is the total acquisition time and $T(t)$ represents the transformation at each time point; it should be noted that voxel i' in $a_{i'j}$ may not correspond to the same voxel i after applying $T(t)$.

2.3. Study of Optimal Conditions for Awake Acquisition with Point Sources

In this study, we aimed to investigate the detectability of point sources in PET imaging using a rat numerical phantom with four point sources and to investigate how these sources affect brain uptake estimation. The phantom was designed with two point sources positioned at the snout and two under the ear.

To evaluate the detectability of point sources and their impact on brain uptake estimation, we explored the effect of different parameters, including the activity of the numerical rat phantom and the activity of the point sources. Specifically, we varied the activity of the rat brain phantom in steps of 20 µCi , ranging from 10 to 210 µCi, and the activity of the point sources in steps of 0.5 µCi, ranging from 1 to 10 µCi. This resulted in a total of 220 combinations of brain and point source activities.

It is important to note that the brain activity simulated in our experiments corresponds to approximately 15% of the total activity in the rat body. This percentage represents the median activity level observed in the brain across the four rat acquisitions explained in Section 2.4. Since the process involves stochastic elements, each combination was simulated 100 times, randomly moving the rat within the FOV to obtain a detectability value for each case.

For each simulation, we used a time step of 12.5 ms (corresponding to a frequency of 80 Hz). This choice of time step strikes a balance between precise motion tracking and good detectability of the sources in the image. The execution time of each simulation was not lengthy due to the short time step. Additionally, to address the introduction of more random coincidences with increasing acquisition activity, we investigated how well the delayed window (DW) method, as proposed by Yavuz et al. [24], can mitigate this issue by subtracting the contribution of random coincidences from the image.

Apart from detectability, we also studied how the activity of the point sources affects image quality. Our primary goal is to study the brain of the animal accurately, which requires avoiding halo artifacts induced by the point sources attached to the animal's head. Halo artifacts are circular regions around high-activity areas, such as the point sources, where nearby regions underestimate the uptake [25]. In addition, we considered the impact of random coincidences introduced by the sources, and we assessed how the DW method can help reduce their impact.

To investigate the effects of point source activity on brain quantification, we conducted simulations with different activities injected into the animal, both with and without point sources. We used the image without point sources as the reference and computed the Root Mean Square Error (RMS_e) for the images with different point source activities. The RMS_e values are defined as

$$RMS_e = \sqrt{\frac{\sum_{j \in BR} (I_j^o - I_j^{ps})^2}{N}} \tag{3}$$

where the sum is performed over the voxels j inside the brain region BR. I^o represents the image without point sources, and I^{ps} represents the image with point sources. N is the total number of voxels inside the brain region. To ensure fair comparison, both images are in relative standardized uptake value (SUV_r).

All simulations took into account scatter and random events to accurately model real-world conditions.

The results of this study enable us to establish the optimal conditions for detecting point sources while avoiding compromising the quality of the reconstructed brain image by adding too much activity.

2.4. Study of Rat Behavior in PET Scanner and How Count Subtraction Affects the Image

The objective of this section is to investigate the effects of rat behavior on PET imaging data, particularly focusing on the impact of subtracting coincidences from the acquisition, as shown in steps 2 and 4 in Figure 1.

To assess the effects of animal behavior on PET imaging data, we conducted a comprehensive study using four different Wistar rats injected with ^{18}F-FDG within the 6R-SuperArgus scanner. During the experiments, the rats were awake. The rats were introduced into a tube that offered freedom of movement. However, the limited diameter of the tube prevented the animals from making full turns, thereby ensuring that they remained within the FOV of the scanner. Acquisitions lasted approximately 600 s for each rat. A primary objective was to identify the parts of the acquisition where the rats moved too quickly, as such movements can adversely affect image quality. Consequently, we subtracted these fastest motion data from the final image reconstruction to improve the overall accuracy.

In one of the rats, four point sources of ^{18}F were placed at the same positions as the simulation shown in Figure 2. This rat is a wistar female rat weighing 255 g. Each point source had an activity of 7 µCi, while the rat's brain had a total activity of 110 µCi at the beginning of the acquisition. Figure 3 shows the rat with the point sources. The study focused on exploring the effects of subtracting more or fewer coincidences by varying the parameters in the reconstruction process. The two parameters that significantly impact the number of counts are v_{max} and χ^2_{max}, as discussed earlier.

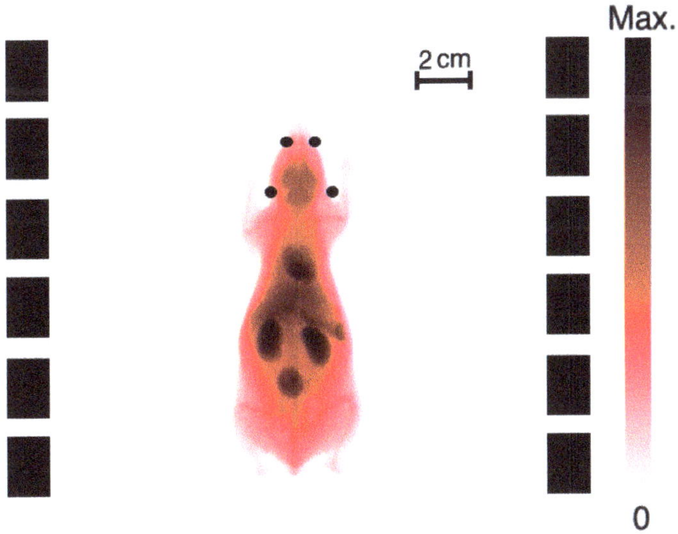

Figure 2. Schematic representation of the numerical rat phantom used in our simulations [26], located inside the 6R-SuperArgus PET scanner. The phantom includes four point sources, two at the snout and two under the ear. The point sources are shown larger than their actual size (1 mm diameter) for visualization purposes.

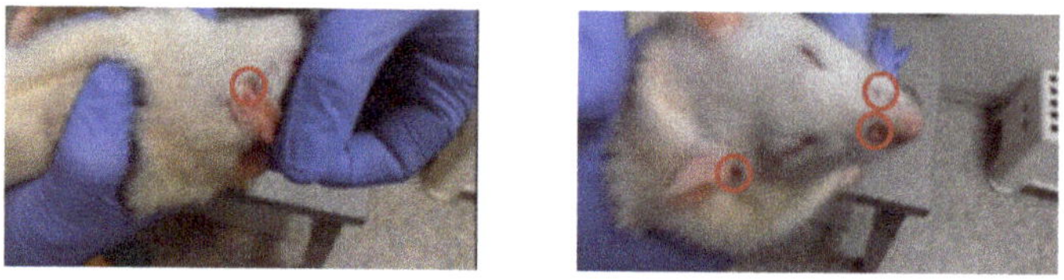

Figure 3. Rat with the point sources attached to the head.

Two different metrics were used to assess the effect of these parameters and the subtraction of counts. First, we used the point sources attached to the animal's brain to measure the precision of motion compensation by calculating the Full Width at Half Maximum (FWHM) of these sources. The FWHM is a useful measure of spatial resolution that allows us to assess the amount of blurring caused by movement during acquisition.

Secondly, we used the cortex region to measure the signal-to-noise ratio (SNR) , which provides valuable insight into the impact of count subtraction on image quality. SNR is defined by the following formula:

$$\mathrm{SNR} = \frac{\mu}{std} \quad (4)$$

Here, μ represents the mean value inside the region of the cortex, while *std* denotes the standard deviation within the same region. A higher SNR indicates better image quality with reduced noise.

By analyzing the FWHM and SNR under different conditions, we aim to understand how the movement of rats during PET acquisitions affects image quality and how count subtraction influences the final reconstruction. These insights will contribute to the optimization of point source tracking in Awake Rat PET Imaging, leading to more accurate and reliable data for neuroscience research and other related fields.

2.5. Comparison of Awake vs. Anesthetized Brain Reconstruction

To evaluate the performance of our method, we conducted experiments on a female Wistar rat weighing 255 g, on which four ^{18}F point sources were placed, as detailed in the preceding section. Each point source had an activity of 7 µCi, while the rat's brain had a total activity of 110 µCi at the beginning of the acquisition. The experiments were carried out in two states: under anesthesia and while the rat was awake.

The rat was positioned within the 6R-SuperArgus scanner, initially under anesthesia, with data acquisition commencing just prior to the onset of the awakening process. Each data acquisition session lasted for 600 s. For the anesthetized state, the acquisition yielded a total of 1.55×10^8 coincidences. Once the rat had fully awakened, we performed the second data acquisition, resulting in 1.45×10^8 coincidences.

By comparing the data obtained from the awake and anesthetized states, we gained valuable insights into how motion affected image quality within the context of PET imaging. This comparative analysis enabled us to assess the effectiveness of our motion compensation methods by directly contrasting the resulting images.

3. Results

3.1. Study of Optimal Conditions for Awake Acquisition with Point Sources

The results presented in this section contribute significantly to understanding the relationship between the activity of both the animal and the point sources and the success rate of tracking the point sources. Figure 4 displays the percentage rate of correct tracking of the point sources for frames of 12.5 ms. We have explored how Delayed Window (DW) correction affects detectability, and it is evident that DW correction has a positive impact on the detectability of short frames, expanding the scenarios in which all point sources can be reliably detected. The dash-dotted line serves as a visual representation of the desired activity configuration in an experiment, where there is a 100% rate of correct tracking of the point sources. This figure demonstrates the effectiveness of the tracking system in high-activity scenarios and provides a reference for optimizing future tracking systems. The results presented in this figure are crucial for guiding experimental design and determining the optimal conditions for point source tracking.

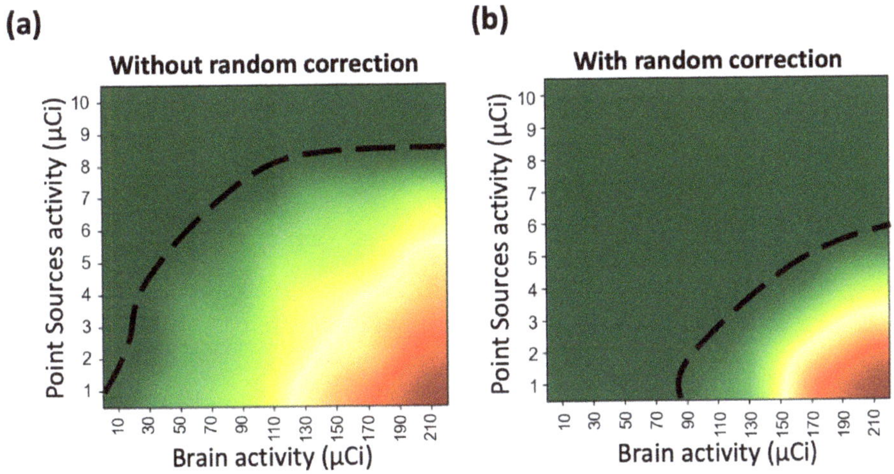

Figure 4. Point source tracking success as a function of animal and point source activity for frames of 12.5 ms: (**a**) without random correction, (**b**) with random correction using the delayed window method. Since the point source locations are known, the success rate is defined as the percentage of time that all four sources are correctly located. The area above the dashed line represents the ideal activity configuration, where all point sources are tracked with 100% accuracy. This figure is the result of 100 simulations for each configuration.

Next, we investigated how the presence of these point sources affects image quality. Figure 5 illustrates the root mean square error (RMS_e) between the image without point sources and the image with point sources at the injected activity of the animal. All simulations encompassed 450 s of acquisition, considering that our acquisitions are of 600 s, and we estimate a loss of counts of approximately 25% due to the methods of subtraction mentioned in Section 2.2. In this case, DW correction is necessary, as it consistently improves the image quality in all cases. The region below the dash line in Figure 5 represents the range of point source activity that has an RMS_e of less than 0.05, which we consider to have a negligible effect on the quantification of activity in different brain regions of the animal.

By combining the studies on detectability and image quality, we can identify the region of optimal conditions for awake acquisitions with point sources. Figure 6 depicts this region, shown in green. These conditions ensure that the point sources can be tracked every 12.5 ms, and the reconstructed image has a lower degradation than 0.05 RMS_e compared to the image without point sources. The star in the figure represents the conditions of the acquisition analyzed in Sections 3.2 and 3.3.

The information presented in this section provides valuable insights into the best conditions for awake PET imaging with point sources. These findings will contribute significantly to the advancement of motion detection and correction techniques in this field and serve as a foundation for further optimizing tracking systems in future experiments.

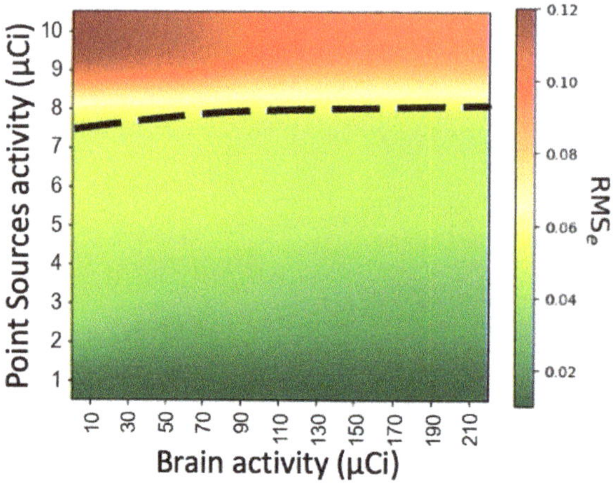

Figure 5. RMS_e as a function of animal and point sources activity for studies with 450 s acquisitions. The region below the dashed line represents the activity configuration that has an RMS_e of less than 0.05. All images have DW correction.

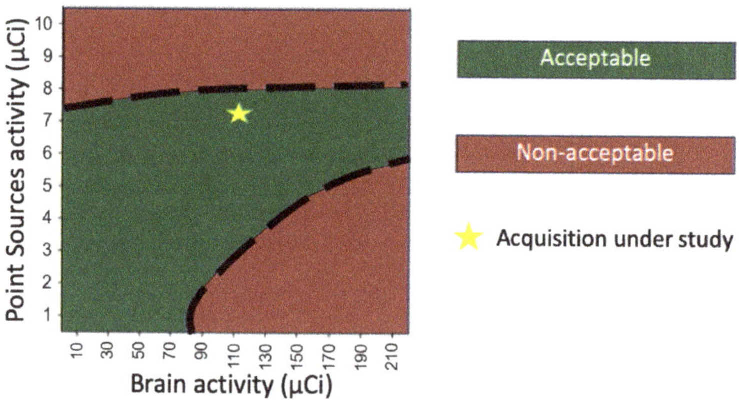

Figure 6. Regions of acceptable and non-acceptable conditions for awake acquisitions as a function of animal and point source activity. The green region represents the area where the sources can be tracked every 12.5 ms, and the image reconstructed has a lower degradation than 0.05 RMS_e with respect to the image without point sources. The star represents the conditions of the acquisition analyzed in Sections 3.2 and 3.3.

3.2. Study of Rat Behaviour in PET Scanner and How Count Subtraction Affects the Image

In this section, our aim is to understand the effect of subtracting counts from the original acquisition. We focus on the subtraction of counts during quick-movement phases. As mentioned before, in order to achieve higher tracking success, we avoid coincidences where the animal is moving quickly, but a trade off between better tracking and count loss is at play. We also investigate the behavior of four different rats inside the 6R-SuperArgus scanner while moving freely in order to identify fast motion periods.

Figure 7 displays the study of the movement of four rats inside the 6R-SuperArgus scanner, showing the centroid of LORs every 50 ms. The areas in green represent regions with low movement, while those in red indicate areas categorized as quick movement. At the top of each graphic, the percentage of events inside low movement frames is shown.

We conducted tests with four different animals to assess the pattern of rat motion inside the scanner and how many events would be removed in our approach. Figure 7 reveals that we retain an average of 73.75% of the counts, with count subtractions ranging from 13.5% in the case with the lowest animal movement to 43.3% in the worst case. We can adjust the v_{max} value to achieve the best reconstruction, as shown shortly.

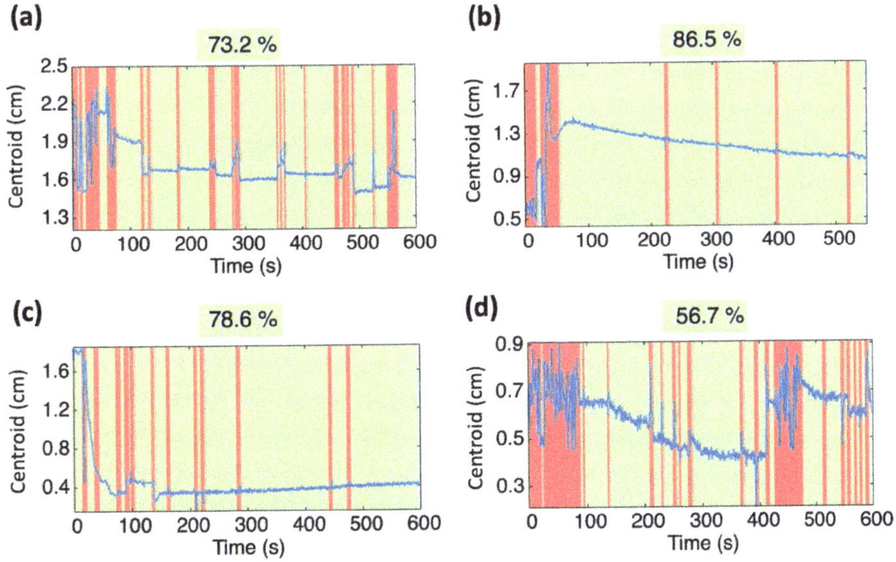

Figure 7. Study of the movement of four rats inside a 6R-SuperArgus scanner. Blue lines show the centroid of LORs every 50 ms. Green areas indicate regions with low movement, while red areas represent regions categorized as quick movement. The percentage of events inside low movement frames is shown at the top of each graphic.

All studies have a total acquisition time of 600 s, providing a sufficient number of events to obtain noise-free images despite possible statistical loss. Additionally, it can be observed that, in most cases, at the early stages of the acquisition, the animal displays significant movement but, after a brief period, relaxes and reduces the amount of movement over time.

Now, we focus on the rat located in the top-left quadrant of Figure 7, which features four point sources attached to its head. The point and rat activities were chosen to lie in the region of optimal conditions, marked with a star in Figure 6. Now, further, we have to adjust two key parameters: v_{max}, which controls the acceptance range for the speed of movement, and χ^2_{max}, which governs the tolerance for accepting less accurate point source position determination. Exploring these parameters enables us to understand the trade off between accepting more or fewer counts.

In Figure 8, we present a comprehensive overview of our study. Panel a shows how higher tolerances in both v_{max} and χ^2_{max} result in keeping more counts in the reconstruction. Panel b shows the trade off between the number of counts used and the apparent size of the reconstructed point sources. This panel suggests that the optimal choice for v_{max} is 1.0 mm/s, as, across different χ^2_{max} values, deviating from this value increases the apparent FWHM of the sources. Additionally, when χ^2_{max} exceeds 0.055, we observe a deterioration in resolution.

Panel c showcases the region of interest (ROI) within the cortex, which is utilized to compute SNR values presented in Panel d. We can see that too strict criteria to accept counts result in pronounced noise in the image, leading to a smaller SNR in the cortex region. Conversely, if we accept nearly all counts, as seen in the right-most case in Figure 9, we introduce noise due to poorly compensated motion, ultimately degrading the image. The optimal combination of parameters yielding the best SNR and FWHM values is achieved when χ^2_{max} is close to 0.05 and v_{max} is set to 1 mm/s.

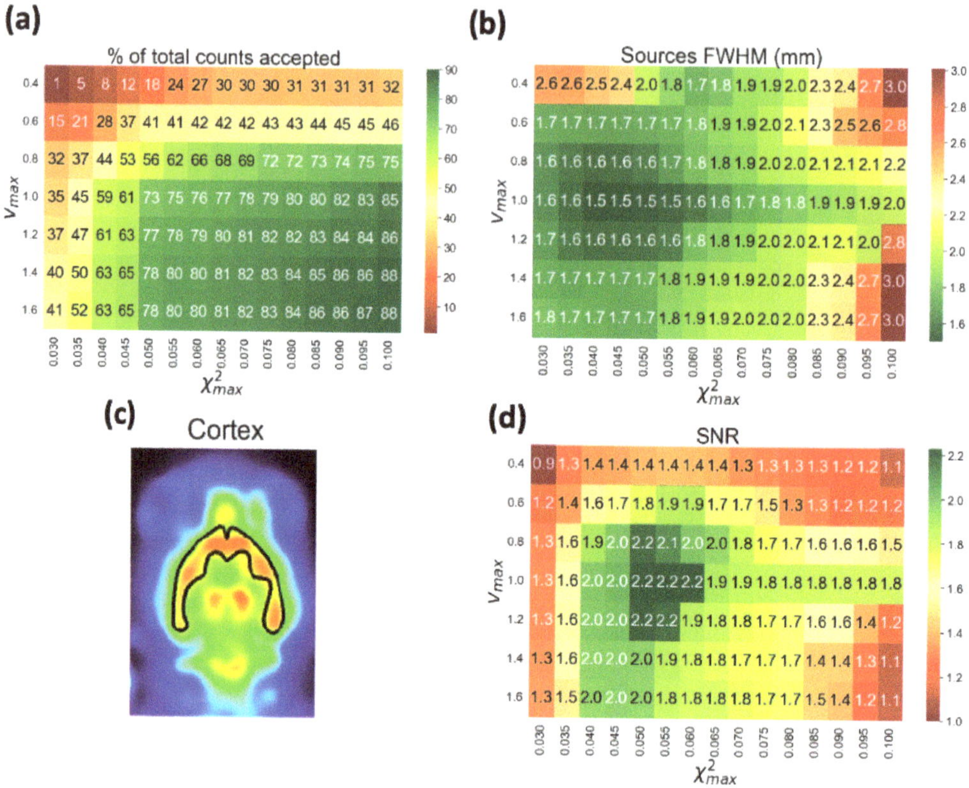

Figure 8. Study on the impact of count subtraction on the image. The total number of coincidences in the acquisitions is 1.45×10^8. (**a**) Percentage of admitted counts for each combination of v_{max} and χ^2_{max}. (**b**) Average FWHM of the four point sources for each case. (**c**) Cortex region utilized for SNR calculation. (**d**) SNR values corresponding to each case.

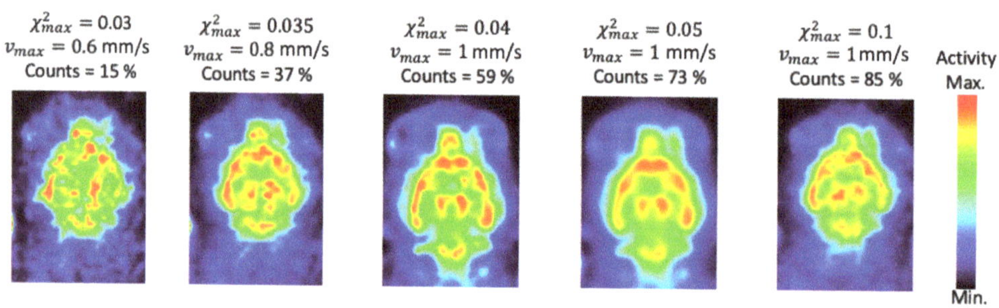

Figure 9. Reconstruction of five different scenarios with varied v_{max} and χ^2_{max}.

3.3. Comparison of Awake vs. Anesthetized Brain Reconstruction

In this section, we compare the imaging results of a rat under anesthesia and in an awake state. The awake state of this acquisition is shown in Figure 7a by its centroid.

In the awake state, 73% of the events were retained after filtering the rapid motion regions, using the best parameters determined in the previous section. Both reconstructions reveal distinguishable brain structures, with only an 8% increase in FWHM of the point sources in the awake rat, with a 6% decrease on SNR at the cortex. Consequently, we can conclude that animal studies can be conducted in awake rats without severely affecting the quantification in brain regions.

4. Discussion

In this study, a primary objective was to optimize the parameters chosen during tracking of point sources in awake rat PET imaging using a motion detection and correction system. We have achieved significant insights that shed light on the factors influencing detectability and image quality in this motion correction strategy.

Initially, we investigated the effects of injected activity on the detectability of point sources attached to the animal. Through extensive simulations, we demonstrated the importance of random correction in ensuring the detection of all point sources. It was also found that that, to obtain image deviations below RMS_e of 0.05 of the reference, requires keeping the activity of the sources below 8 μCi. Additionally, the point sources require a minimum activity when the brain's activity exceeds 90 μCi. While this study was performed using FDG as the tracer, we acknowledge that the activity levels in other organs may vary depending on the tracer used, potentially affecting random coincidences. Therefore, future studies using different tracers should take this into account when optimizing tracking systems.

Next, we addressed the trade off of subtracting coincidences associated with quick-movement and non-precise point source tracking. Through our observations of the four rats within the scanner, as shown in Figure 7, we identified a tendency for these animals to display increased movement during the initial stages of the acquisition. This initial movement could be attributed to the novelty of the environment. However, it became evident that, as the rats acclimated to the scanner, movement reduced.

By optimizing our motion detection and correction parameters, we achieved low noise and excellent image resolution, with the best images obtained when retaining 73% of coincidences for our rat. Remarkably, even with a loss of 27% of coincidences, the impact on image resolution was only 8% and on SNR was 6%, as demonstrated in Figure 10. While our current experiments were conducted with a 10 min acquisition time, future research will explore the potential advantages of longer acquisition durations. Extending the acquisition time may provide an opportunity to capture additional information and enhance imaging sensitivity. However, it is important to recognize that longer acquisition times can introduce challenges related to increased subject movement, necessitating further investigation into the associated motion correction techniques and potential limitations.

We note that the position of the animal's head near the end of the FOV of the scanner may have contributed to the loss of resolution observed. The non-homogeneous resolution across the FOV of the scanner, which exhibits a poorer resolution at the edges of the FOV and could have worsened the image quality. For future studies, it would be beneficial to consider this non-uniform resolution when placing the animal and optimizing motion detection and correction systems for awake rat PET imaging. Additionally, exploring methods to improve the resolution near the edges of the FOV could further enhance image quality in awake rat PET studies [27].

Another important aspect is that, recently, Miranda et al. [28] have proposed adding corrections for cases where point sources shift on the animal's skin. In our approach, instead of trying to correct for these, we remove them from the acquisition by introducing the χ^2 parameter. This way, whether the sources have shifted on the animal or an incorrect transformation has been computed, these counts will not introduce erroneous information into the reconstruction. The loss of counts introduced this way is moderate and can be compensated by a modest increase in acquisition time.

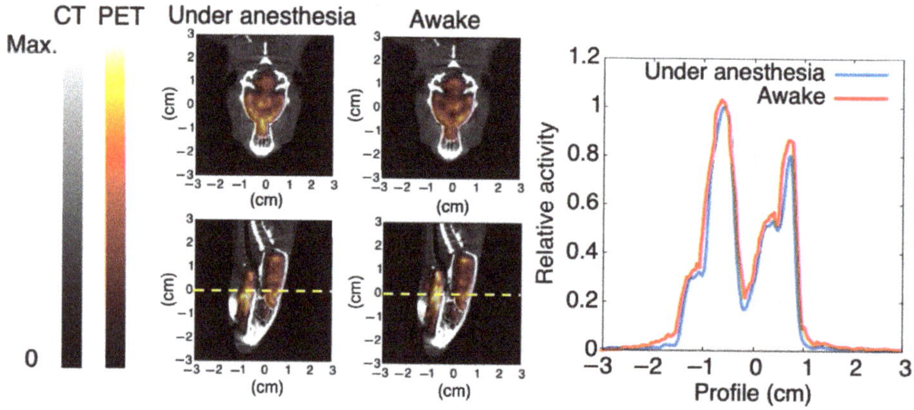

Figure 10. Rat head study with registered CT image in the 6R-SuperArgus scanner. On the **left**, sagittal and coronal views of a rat in both states, awake and under anesthesia. On the **right**, the profile of the yellow dashed line is shown.

5. Conclusions

This study offers valuable insights to optimize the parameters for ^{18}F point source tracking in awake rat PET imaging and establishes a methodology for determining the appropriate marker activity levels relative to rat-injected activity. These levels are scanner-dependent, contingent on sensitivity and resolution. Focusing on the 6R-SuperArgus scanner, we found that random corrections are of great importance and that combining these random correction techniques with carefully selected motion detection and correction parameters ensures comparable image quality to anesthetized acquisitions. By selecting coincidences during periods of no-quick rat motion (approximately 70–80% of acquisition time), we can produce high-quality brain images with only minor resolution reduction, yielding minimal disparities compared to anesthetized rat studies. In summary, we demonstrate that appropriately dosed ^{18}F point markers can facilitate motion detection and compensation in awake rat PET studies, emphasizing the importance of tailoring the approach to study-specific conditions for image comparability with sedated rat studies.

Author Contributions: Software (F.A.-V., P.G., J.M.U. and J.L.H.); data and resources (J.J.V. and M.D.); writing and reviewing (all authors). All authors have read and agreed to the published version of the manuscript.

Funding: This work was supported by the Spanish Government (FASCINA PID2021-126998OB-I00, NEWMBI CPP2021-008751, 3PET PDC2022-133057-I00), Comunidad de Madrid (ASAP-CM, S2022/BMD-7434), and the Recovery, Resilience, and Transformation Plan financed by the European Union through Next Generation EU funds. This work was also supported by the Spanish Ministry of Economic Affairs and Digital Transformation (Project MIA.2021.M02.0005 TARTAGLIA, from the Recovery, Resilience, and Transformation Plan financed by the European Union through Next Generation EU funds). TARTAGLIA takes place under the R&D Missions in Artificial Intelligence program, which is part of the Spain Digital 2025 Agenda and the Spanish National Artificial Intelligence Strategy. P. G. also acknowledges support from the Margarita Salas Fellowship, CT18/22 at Complutense University of Madrid, funded by the European Union-Next-GenerationUE funds.

Institutional Review Board Statement: All experimental procedures have been performed in compliance with the European Communities Council Directive 2010/63/EU and submitted for approval by the Institutional Animal Care and Use and Ethics Committee of the Hospital General Universitario Gregorio Marañón (HGUGM), supervised by the Comunidad de Madrid according to the Annex X of the RD 53/2013. The subjects were kept at the animal housing facilities of the Unidad de Medicina y Cirugía Experimental (UMCE-HGUGM) in Madrid, Spain.

Informed Consent Statement: Not applicable.

Data Availability Statement: The data that support the findings of this study are available from the corresponding author upon reasonable request.

Acknowledgments: We thank C. G. Fernandez, J. M. Arco, R. Matesanz, and J. López-Longas from the SEDECAL company for granting access to the reference machine data and providing experimental support. We also thank Alejandra de Francisco, Yolanda Sierra, and María de la Jara Felipe from the Laboratory of Medical Imaging (LIM) for the handling of the animals.

Conflicts of Interest: The authors declare no conflict of interest.

References

1. Alstrup, A.K.; Landau, A.M.; Holden, J.E.; Jakobsen, S.; Schacht, A.C.; Audrain, H.; Wegener, G.; Hansen, A.K.; Gjedde, A.; Doudet, D.J. Effects of anesthesia and species on the uptake or binding of radioligands in vivo in the Göttingen minipig. *Biomed Res. Int.* **2013**, *2013*, 808713. [CrossRef] [PubMed]
2. Toyama, H.; Ichise, M.; Liow, J.S.; Vines, D.C.; Seneca, N.M.; Modell, K.J.; Seidel, J.; Green, M.V.; Innis, R.B. Evaluation of anesthesia effects on [^{18}F] FDG uptake in mouse brain and heart using small animal PET. *Nucl. Med. Biol.* **2004**, *31*, 251–256. [CrossRef] [PubMed]
3. Miranda, A.; Bertoglio, D.; Stroobants, S.; Staelens, S.; Verhaeghe, J. Translation of preclinical PET imaging findings: Challenges and motion correction to overcome the confounding effect of anesthetics. *Front. Med.* **2021**, *8*, 753977. [CrossRef] [PubMed]
4. Alstrup, A.K.O.; Smith, D.F. Anaesthesia for positron emission tomography scanning of animal brains. *Lab. Anim.* **2013**, *47*, 12–18. [CrossRef] [PubMed]
5. Patel, V.D.; Lee, D.E.; Alexoff, D.L.; Dewey, S.L.; Schiffer, W.K. Imaging dopamine release with positron emission tomography (PET) and 11C-raclopride in freely moving animals. *Neuroimage* **2008**, *41*, 1051–1066. [CrossRef] [PubMed]
6. Sung, K.K.; Jang, D.P.; Lee, S.; Kim, M.; Lee, S.Y.; Kim, Y.B.; Park, C.W.; Cho, Z.H. Neural responses in rat brain during acute immobilization stress: A [F-18] FDG micro PET imaging study. *Neuroimage* **2009**, *44*, 1074–1080. [CrossRef] [PubMed]
7. Hosoi, R.; Matsumura, A.; Mizokawa, S.; Tanaka, M.; Nakamura, F.; Kobayashi, K.; Watanabe, Y.; Inoue, O. MicroPET detection of enhanced 18F-FDG utilization by PKA inhibitor in awake rat brain. *Brain Res.* **2005**, *1039*, 199–202. [CrossRef]
8. Momosaki, S.; Hatano, K.; Kawasumi, Y.; Kato, T.; Hosoi, R.; Kobayashi, K.; Inoue, O.; Ito, K. Rat-PET study without anesthesia: Anesthetics modify the dopamine D1 receptor binding in rat brain. *Synapse* **2004**, *54*, 207–213. [CrossRef]
9. Suzuki, C.; Kosugi, M.; Magata, Y. Conscious rat PET imaging with soft immobilization for quantitation of brain functions: Comprehensive assessment of anesthesia effects on cerebral blood flow and metabolism. *EJNMMI Res.* **2021**, *11*, 46. [CrossRef]
10. Schulz, D.; Southekal, S.; Junnarkar, S.S.; Pratte, J.F.; Purschke, M.L.; Stoll, S.P.; Ravindranath, B.; Maramraju, S.H.; Krishnamoorthy, S.; Henn, F.A.; et al. Simultaneous assessment of rodent behavior and neurochemistry using a miniature positron emission tomograph. *Nat. Methods* **2011**, *8*, 347–352. [CrossRef]
11. Woody, C.; Schlyer, D.; Vaska, P.; Tomasi, D.; Solis-Najera, S.; Rooney, W.; Pratte, J.F.; Junnarkar, S.; Stoll, S.; Master, Z.; et al. Preliminary studies of a simultaneous PET/MRI scanner based on the RatCAP small animal tomograph. *Nucl. Instruments Methods Phys. Res. Sect. Accel. Spectrometers, Detect. Assoc. Equip.* **2007**, *571*, 102–105. [CrossRef]
12. Woody, C.; Kriplani, A.; O'connor, P.; Pratte, J.F.; Radeka, V.; Rescia, S.; Schlyer, D.; Shokouhi, S.; Stoll, S.; Vaska, P.; et al. RatCAP: A small, head-mounted PET tomograph for imaging the brain of an awake RAT. *Nucl. Instruments Methods Phys. Res. Sect. Accel. Spectrometers, Detect. Assoc. Equip.* **2004**, *527*, 166–170. [CrossRef]
13. Vaska, P.; Woody, C.; Schlyer, D.; Pratte, J.F.; Junnarkar, S.; Southekal, S.; Stoll, S.; Schulz, D.; Schiffer, W.; Alexoff, D.; et al. The design and performance of the 2 nd-generation RatCAP awake rat brain PET system. In Proceedings of the 2007 IEEE Nuclear Science Symposium Conference Record, Honolulu, HI, USA, 26 October–3 November 2007; Volume 6, pp. 4181–4184.
14. Kyme, A.Z.; Zhou, V.; Meikle, S.R.; Fulton, R.R. Real-time 3D motion tracking for small animal brain PET. *Phys. Med. Biol.* **2008**, *53*, 2651. [CrossRef] [PubMed]
15. Kyme, A.Z.; Zhou, V.W.; Meikle, S.R.; Baldock, C.; Fulton, R.R. Optimised motion tracking for positron emission tomography studies of brain function in awake rats. *PLoS ONE* **2011**, *6*, e21727. [CrossRef] [PubMed]
16. Spangler-Bickell, M.G.; de Laat, B.; Fulton, R.; Bormans, G.; Nuyts, J. The effect of isoflurane on 18F-FDG uptake in the rat brain: A fully conscious dynamic PET study using motion compensation. *EJNMMI Res.* **2016**, *6*, 1–10. [CrossRef] [PubMed]
17. Miranda, A.; Staelens, S.; Stroobants, S.; Verhaeghe, J. Estimation of and correction for finite motion sampling errors in small animal PET rigid motion correction. *Med. Biol. Eng. Comput.* **2019**, *57*, 505–518. [CrossRef] [PubMed]
18. Torheim, T.; Malinen, E.; Kvaal, K.; Lyng, H.; Indahl, U.G.; Andersen, E.K.; Futsaether, C.M. Classification of dynamic contrast enhanced MR images of cervical cancers using texture analysis and support vector machines. *IEEE Trans. Med. Imaging* **2014**, *33*, 1648–1656. [CrossRef]
19. Miranda, A.; Staelens, S.; Stroobants, S.; Verhaeghe, J. Markerless rat head motion tracking using structured light for brain PET imaging of unrestrained awake small animals. *Phys. Med. Biol.* **2017**, *62*, 1744. [CrossRef]
20. Miranda, A.; Staelens, S.; Stroobants, S.; Verhaeghe, J. Fast and accurate rat head motion tracking with point sources for awake brain PET. *IEEE Trans. Med. Imaging* **2017**, *36*, 1573–1582. [CrossRef]
21. Miranda, A.; Glorie, D.; Bertoglio, D.; Vleugels, J.; De Bruyne, G.; Stroobants, S.; Staelens, S.; Verhaeghe, J. Awake 18F-FDG PET imaging of memantine-induced brain activation and test–retest in freely running mice. *J. Nucl. Med.* **2019**, *60*, 844–850. [CrossRef]

22. Arias-Valcayo, F.; Herraiz, J.L.; Galve, P.; Vaquero, J.; Desco, M.; Udías, J.M. Awake preclinical brain PET imaging based on point sources. In Proceedings of the 15th International Meeting on Fully Three-Dimensional Image Reconstruction in Radiology and Nuclear Medicine, SPIE, Philadelphia, PA, USA, 2–6 June 2019; Volume 11072, pp. 546–550.
23. Udias, J.; Gutierrez Fernandez, C.; Herraiz, J.; Perez-Benito, D.; Galve, P.; Lopez-Montes, A.; Lopez-Longas, J.; Arco, J.; Desco, M.; Vaquero, J. Performance evaluation of the PET subsystem of the extended FOV SuperArgus 6R preclinical scanner. In Proceedings of the IEEE Nuclear Science Symposium and Medical Imaging Conference, Sydney, Australia, 10–17 November 2018; pp. 10–11.
24. Yavuz, M.; Fessler, J.A. Statistical image reconstruction methods for randoms-precorrected PET scans. *Med. Image Anal.* **1998**, *2*, 369–378. [CrossRef]
25. Heußer, T.; Mann, P.; Rank, C.M.; Schäfer, M.; Dimitrakopoulou-Strauss, A.; Schlemmer, H.P.; Hadaschik, B.A.; Kopka, K.; Bachert, P.; Kachelrieß, M.; et al. Investigation of the halo-artifact in 68Ga-PSMA-11-PET/MRI. *PLoS ONE* **2017**, *12*, e0183329. [CrossRef]
26. Galve, P.; Arias-Valcayo, F.; Montes, A.; Villa-Abaunza, A.; Ibanez, P.; Herraiz, J.; Udías, J. Ultra-fast Monte Carlo PET Reconstructor. In Proceedings of the 16th International Meeting on Fully 3D Image Reconstruction in Radiology and Nuclear Medicine, Stony Brook, NY, USA, 16–21 July 2021; pp. 152–156.
27. Arias-Valcayo, F.; Galve, P.; Herraiz, J.L.; Desco, M.; Vaquero, J.J.; Udías, J.M. Reconstruction of multi-animal PET acquisitions with anisotropically variant PSF. *Biomed. Phys. Eng. Express* **2023**, *9*, 065018. [CrossRef]
28. Miranda, A.; Kroll, T.; Schweda, V.; Staelens, S.; Verhaeghe, J. Correction of motion tracking errors for PET head rigid motion correction. *Phys. Med. Biol.* **2023**, *68*, 175009. [CrossRef]

applied sciences

MDPI

Article

MDAU-Net: A Liver and Liver Tumor Segmentation Method Combining an Attention Mechanism and Multi-Scale Features

Jinlin Ma [1,2], Mingge Xia [1,*], Ziping Ma [3] and Zhiqing Jiu [1]

1 School of Computer Science and Engineering, North Minzu University, Yinchuan 750021, China; majinlin@nmu.edu.cn (J.M.); j2313795280@163.com (Z.J.)
2 Key Laboratory of Images and Graphics Intelligent Processing of National Ethnic Affairs Commission, North Minzu University, Yinchuan 750021, China
3 School of Mathematics and Information Science, North Minzu University, Yinchuan 750021, China; 2006041@nmu.edu.cn
* Correspondence: xiaamingge@163.com

Abstract: In recent years, U-Net and its extended variants have made remarkable progress in the realm of liver and liver tumor segmentation. However, the limitations of single-path convolutional operations have hindered the full exploitation of valuable features and restricted their mobility within networks. Moreover, the semantic gap between shallow and deep features proves that a simplistic shortcut is not enough. To address these issues and realize automatic liver and tumor area segmentation in CT images, we introduced the multi-scale feature fusion with dense connections and an attention mechanism segmentation method (MDAU-Net). This network leverages the multi-head attention (MHA) mechanism and multi-scale feature fusion. First, we introduced a double-flow linear pooling enhancement unit to optimize the fusion of deep and shallow features while mitigating the semantic gap between them. Subsequently, we proposed a cascaded adaptive feature extraction unit, combining attention mechanisms with a series of dense connections to capture valuable information and encourage feature reuse. Additionally, we designed a cross-level information interaction mechanism utilizing bidirectional residual connections to address the issue of forgetting a priori knowledge during training. Finally, we assessed MDAU-Net's performance on the LiTS and SLiver07 datasets. The experimental results demonstrated that MDAU-Net is well-suited for liver and tumor segmentation tasks, outperforming existing widely used methods in terms of robustness and accuracy.

Keywords: semantic segmentation; liver tumor; attention mechanism; feature fusion; U-Net

Citation: Ma, J.; Xia, M.; Ma, Z.; Jiu, Z. MDAU-Net: A Liver and Liver Tumor Segmentation Method Combining an Attention Mechanism and Multi-Scale Features. *Appl. Sci.* **2023**, *13*, 10443. https://doi.org/10.3390/app131810443

Academic Editors: Ioannis A. Kakadiaris, Michalis Vrigkas and Christophoros Nikou

Received: 15 June 2023
Revised: 8 September 2023
Accepted: 11 September 2023
Published: 18 September 2023

1. Introduction

The liver is a crucial organ in the metabolic process of the human organism, and liver tumors, as a highly prevalent disease, seriously threaten human life and health. The accurate segmentation of tumor regions from computed tomography (CT) images is an important step in the subsequent diagnostic and therapeutic phases. This process can provide doctors with more precise information about the location of lesions, enhancing diagnostic efficiency and accuracy and offering higher clinical value.

However, it is challenging to effectively and precisely distinguish tumor areas from the background due to the diversity of tumor shapes and locations. In recent years, deep learning methods have progressively taken center stage in the segmentation of liver tumors [1]. Among them, the U-Net [2] model has proven to have a high level of segmentation capabilities. To deal with difficult segmentation tasks, scientists have created numerous U-Net variant networks.

Dickson et al. [3] proposed DCMC-Unet based on two-channel multi-scale convolution for liver tumor segmentation. Meanwhile, they employed a thresholding method to eliminate extraneous tissues for noise elimination. The network can effectively extract features at multiples scales and is applicable to tumors of varying sizes and shapes. However, due to the oversimplified jump connections, the model's ability to facilitate interaction between shallow and deep information is limited. Additionally, the presence of semantic gaps reduces the model's capacity for extracting and combining features. To address this problem, Sabir et al. [4] designed the deep dense network ResU-Net. This replaced the convolutional layers with residual blocks, aiming to make full use of the advantages of the U-Net network and deep residual learning for liver tumor segmentation. Deng et al. [5] introduced deep jump connections into U-Net to fully extract features from the encoder for enhanced feature learning. While the bottleneck layer in the aforementioned two methods is relatively simplistic, encoder features cannot be fully utilized here. This limitation can result in the loss of useful information and the degradation of network performance.

Therefore, to solve the issues mentioned above, we introduced multi-scale feature extraction with dense connections and an attention mechanism U-Net (MDAU-Net) for liver and liver tumor segmentation. The main contributions of this work are as follows:

1. We redesigned the jump connection and introduced a double-flow linear pooling enhancement unit (DLE) to improve the interaction ability between deep and shallow features, which helped to narrow the semantic gap.
2. To better realize the extraction and reuse of useful features, we proposed a cascaded adaptive feature extraction unit (CAE) as a substitute for the bottleneck layer. It was based on an multi-head attention mechanism and a series of dense connections.
3. We designed a cross-level information interaction mechanism (CII). It used bidirectional residual connections and was placed in the skip connection to overcome the problem of forgetting a priori knowledge in the learning process.
4. We proposed a residual encoder to bolster the preservation of original features and supply additional initial information for the segmentation task.

2. Related Works

2.1. Medical Image Segmentation Methods

Medical image segmentation is one of the most important tasks in the field of medical image analysis, aiming to extract quantitative information about various tissue structures and lesions from complicated medical images.

The conventional manual segmentation method utilized in clinical practice entails experienced clinicians manually segmenting raw CT images. This process is characterized by its time-consuming and labor-intensive nature, and the quality of segmentation largely hinges on the operator's experience and medical knowledge. As medical image processing technology has evolved, semi-automatic segmentation methods have gained prominence. These methods encompass thresholding, region growth, statistics, and other automatic segmentation approaches, with deep learning being a prominent representative.

The thresholding method separates the target liver region from the background by selecting the appropriate gray value as the threshold. Seong et al. [6] used a combination of adaptive thresholding and the angular line method to enhance segmentation performance. However, this method is not effective in segmentation when the gray value of the target region is much smaller than the background gray value. The region-growing method initially selects suitable pixel points (i.e., seed points) within the region as the starting point for growth. It then continually adds pixel points with similar properties to achieve segmentation. Chen et al. [7] proposed an automatic liver segmentation method based on the region-growing algorithm. They introduced center-of-mass detection and intensity analysis to ensure quick and accurate liver region segmentation. Additionally, the texture-based region growing method [8] achieves liver segmentation by automatically selecting seed points and calculating a threshold for the region-growth-stopping condition. The selection of seed point locations significantly impacts the performance of the algorithm. Statistics-based segmentation methods [9] demand extensive clinical data as support, limiting the models' generalization ability for small-scale datasets like medical images.

The concept of deep learning was initially introduced by Hinton. In contrast to the traditional methods mentioned above, deep-learning-based methods can automatically learn feature representations from raw data. This significantly enhances the segmentation performance and generalization ability of a model. U-Net is a classic segmentation network in medical image segmentation. Along with its variants, it is widely employed in segmentation tasks due to its low parameter count and superior segmentation performance. Zhou et al. [10] introduced a series of nested dense hopping connections between the encoder and the decoder to enhance U-Net. This resulted in a 3.9% improvement in the mean intersection over union (mean IOU). Huang et al. [11] proposed UNet3+, which made more comprehensive use of the multi-scale features in the feature map. Bi et al. [12] introduced ResCEAttUnet to enhance the network's capacity for extracting multi-scale features. This ensured that the network could effectively capture high-level semantic information while minimizing information loss. Kushnure et al. [13] developed HFRU-Net to meticulously characterize contextual information by local feature reconstruction and feature fusion mechanisms. They also adaptively recalibrated the fused features to emphasize image details. Zhou et al. [14] introduced MCFA-UNet, a multi-scale cascaded feature attention network, to address the issue of edge detail loss resulting from inadequate feature extraction.

2.2. *Atrous Spatial Pyramid Pooling*

As the number of network layers deepens, the resolution of the images decreases, and the generated semantic features become less effective in dense prediction tasks. To tackle this issue, various solutions have been proposed [15–18].

DeepLab V3 [18] is a semantic segmentation model based on atrous convolution, incorporating the atrous spatial pyramid pooling (ASPP) method to effectively fuse features at different scales. As depicted in Figure 1, ASPP comprises five parallel branches: a 1×1 convolutional branch; three atrous convolutional branches with varying expansion rates (6, 12, 18); and a global average pooling branch. The global average pooling branch downsamples the feature maps to a 1×1 size and subsequently upsamples them to the original size using 1×1 convolution and bilinear interpolation. The outputs of these five branches are concatenated to create a richer feature representation. Finally, a 1×1 convolution layer is employed to reduce the number of channels in the feature map to the desired level.

The inclusion of the ASPP structure during liver and liver tumor segmentation can combine the advantages of atrous convolution to expand the receptive field of the convolution kernel without losing resolution. This assists the network in learning semantic information from the multi-scale receptive field, ultimately enhancing the model's segmentation performance.

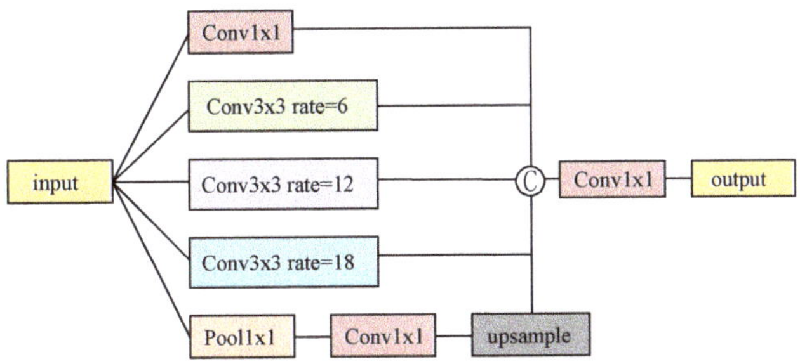

Figure 1. The structure of atrous spatial pyramid pooling.

2.3. *Multi-Head Attention Mechanism*

Attention mechanisms have multiple applications in computer vision as crucial components of neural networks [19,20]. When integrated into the liver tumor segmentation process, attention mechanisms enable the model to adaptively extract lesion features while suppressing irrelevant regions. This ensures that the network focuses on pertinent information for a specific segmentation task.

A generalized attention mechanism can be defined as a method for mapping a query (Q) to a set of keys (K) and values (V). In contrast, the multi-head attention mechanism (MHA) [21] implements a method for mapping a query to multiple key–value pairs. Figure 2 illustrates the structure of the multi-head attention mechanism.

In the multi-head attention mechanism, the input data consist of Q, K, and V matrices. They are mapped to different subspaces by linear transformation to obtain new matrices: $Q_i \in R^{m \times d_k}$, $K_i \in R^{m \times d_k}$, and $V_i \in R^{m \times d_V}$. This transformation is achieved by multiplying them with a learnable weight matrix, as shown in Equation (1).

$$Q_i, K_i, V_i = QW_i^Q, KW_i^K, VW_i^V \quad (1)$$

where W_i^Q, W_i^K and W_i^V denote the learnable weights of the corresponding Q, K, and V, respectively.

Then, scaled dot-product attention is executed for each attention head. This operation is used to compute the attention weights, as shown in Equation (2). It calculates the dot production of Q and K^T to determine the degree of the relationship between Q and K. Subsequently, the outcome is rescaled, and the similarity scores undergo normalization via the softmax function. This process guarantees that the sum of attention weights across all positions equals 1. These weights are employed in a multiplication operation with V to derive the output of the respective attention head.

$$head = Attention(Q, K, V) = softmax\left(\left(QK^T\right) / \left(\sqrt{d_k}\right)\right)V \quad (2)$$

where $\sqrt{d_k}$ represents the scaling factor, which is designed to prevent gradient explosion in the similarity score matrix caused by excessive dimensionality.

Finally, the outputs of each attention head are concatenated and mapped again to obtain the final attention output, as shown in Equation (3).

$$MultiHead(Q, K, V) = (Concatenate(head_1 \cdots head_h))W^o \quad (3)$$

where $head_i$ represents the ith attention head, and the inclusion of multiple attention heads enables the model to concurrently focus on various pieces of subspace information from distinct locations. Additionally, W^o signifies the trainable weight matrix used for linear mapping.

Incorporating multi-head attention into liver and liver tumor segmentation enables the model to selectively extract pertinent features while concurrently attenuating superfluous regions. This strategy guarantees the network's concentration on pertinent information for a given segmentation task, thereby mitigating segmentation errors induced by noisy signals. Furthermore, leveraging multi-head attention empowers the model to enhance its spatial perception, subsequently elevating segmentation accuracy.

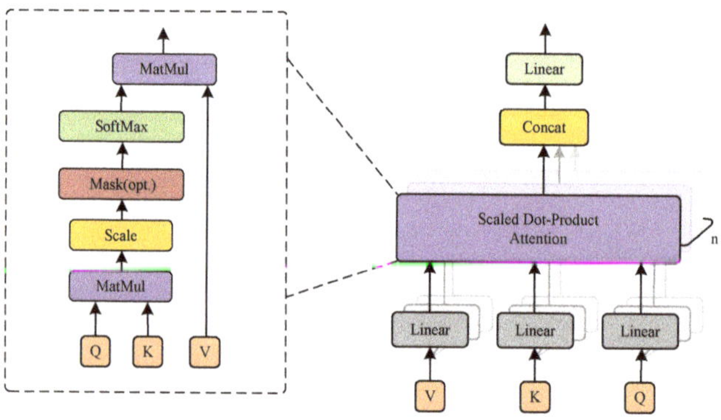

Figure 2. The structure of multi-head attention and scaled dot-product attention.

3. Proposed Method

3.1. Overall Architecture

As shown in Figure 3, MDAU-Net maintains the U-shaped architecture and retains U-Net's decoder path. In contrast to U-Net, MDAU-Net has four key improvements.

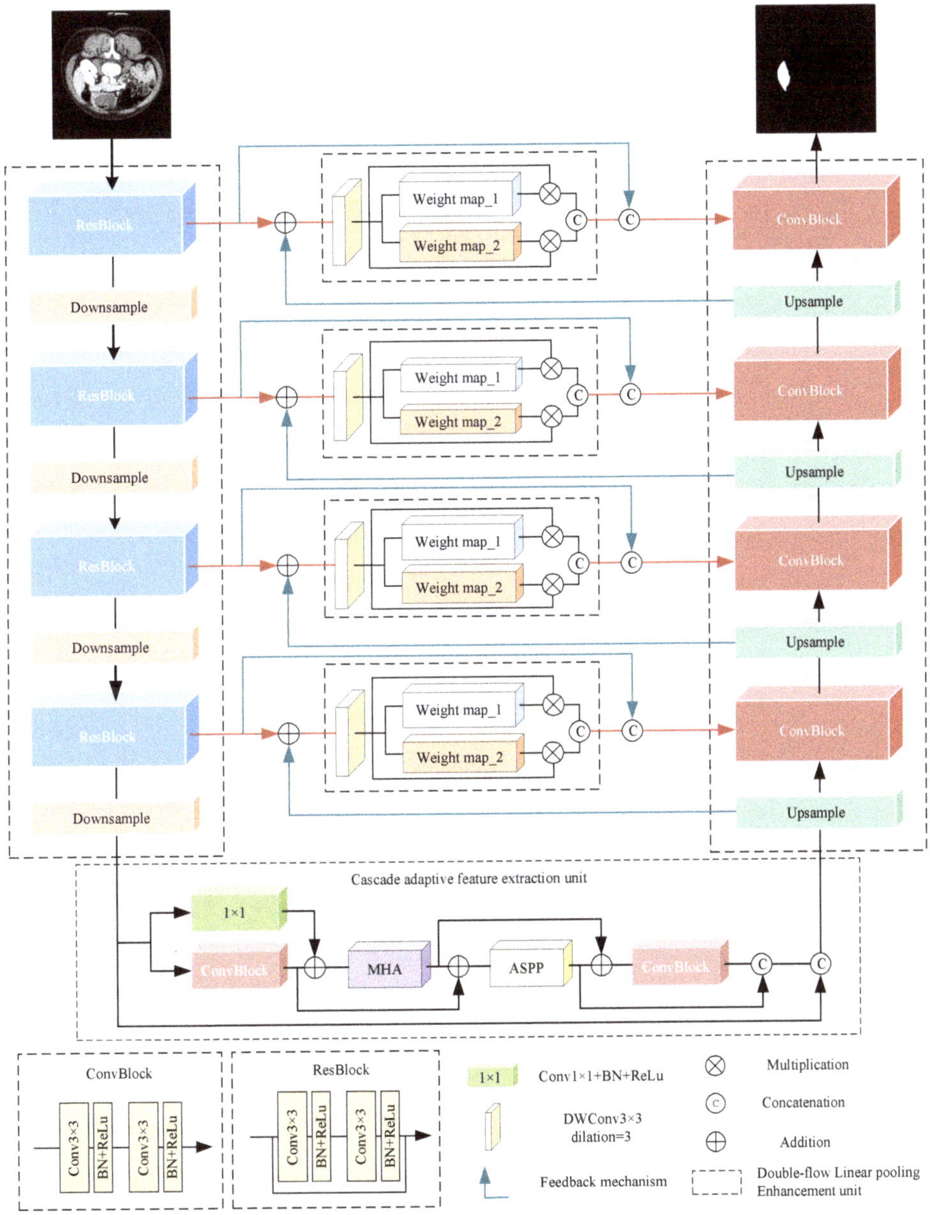

Figure 3. The structure of MDAU-Net.

Firstly, we redesigned the encoder structure. In the original U-Net, the basic block utilizes the ConvBlock structure depicted in Figure 3. However, in MDAU-Net, we incorporated residual connections into the basic block, resulting in the residual encoder. This modification bolstered the preservation of original features and supplied additional initial information for the segmentation task.

Subsequently, to amplify information flow within the network and promote feature reuse, we substituted U-Net's bottleneck layer with a cascaded adaptive feature extraction unit (CAE).

Additionally, we introduced a double-flow linear pooling enhancement unit (DLE) in the jump connection segment to narrow the semantic gap between deep and shallow features through a "progressive" feature fusion approach. This refinement aided the network in achieving more precise target area localization.

Finally, we designed a cross-level information interaction mechanism (CII) utilizing bidirectional residual connections to address the issue of forgetting a priori knowledge during the training process.

In Algorithm 1, we provide a pseudocode as an initial description of MDAU-Net, with a comprehensive exposition of the network's structure to follow in subsequent sections.

Algorithm 1: MDAU-Net

Data: Dataset X, mask L, module parameters
Result: Segmentation result Y

for $i = 1$ *to* N **do**
  Preprocessing and enhancement of image X_i.
  for $j = 1$ *to* 4 **do**
    Encode X_i as E_{ij} using ResBlock and MaxPooling.
    Obtain the feature map E_{ij} for each encoder layer.
  end
  Adaptive feature extraction by CAE module, obtain C_i.
  for $k = 1$ *to* 4 **do**
    Calculate the DLE by E_{ij} and $D_{i(k-1)}$, obtain the feature map T_{ik}.
    Decode C_i as D_{ik} using bilinear interpolation and ConvBlock.
    Obtain the feature map D_{ik} for each decoder layer.
    Obtain the segmentation result Y_i of image X_i as $Y_i = D_{i4}$.
  end
end
Output the segmentation result $Y = [Y_1, Y_2, \ldots, Y_N]$.

3.2. Residual Encoder

The residual structure [22], denoted as ResBlock and introduced as a solution to the gradient vanishing problem, is illustrated in Figure 3. In the encoder path, the repetitive downsampling operation often leads to information loss. Therefore, this study employed a sequence of consecutive residual blocks in lieu of the initial convolutional layer. This approach enhanced the network's capacity to preserve and extract input features effectively. Furthermore, the integration of residual blocks served to mitigate to some degree the gradient vanishing challenge arising from the network's increased depth.

3.3. Cascaded Adaptive Feature Extraction Unit

A single convolutional operation hampers the effective utilization of valuable features in deep networks. In line with the concept of dense connectivity [23], we redefined the bottleneck layer and introduced the cascaded adaptive feature extraction unit (CAE) to facilitate feature reuse and enhance the propagation of useful features throughout the network. Figure 4 illustrates the structure of the CAE unit, which consisted of two convolutional units (Conv_Unit1 and Conv_Unit2), multi-head attention, and atrous spatial pyramid pooling (ASPP). These submodules were interconnected through dense short connections, enabling each module to extract semantic information from the preceding layer or layers, thereby promoting feature reuse and transfer. Additionally, this connectivity aided in the network's convergence.

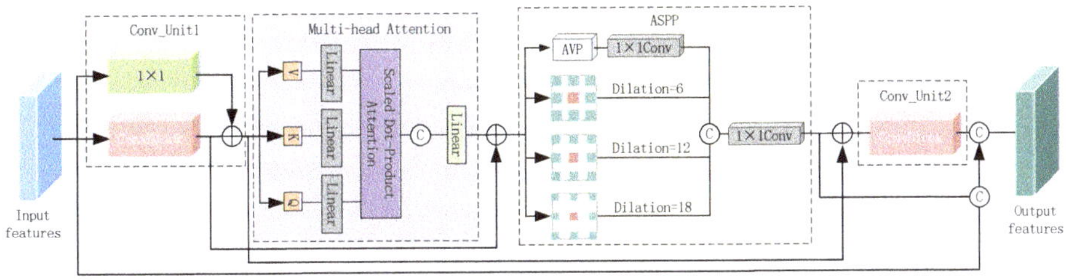

Figure 4. The structure of the cascaded adaptive feature extraction unit (CAE).

Among these submodules, Conv_Unit1 played an essential role in initially enhancing the network's feature representation. Multi-head attention enabled the model to adaptively extract crucial semantic features, focusing on valuable features relevant to liver tumors while reducing the impact of redundant features or background noise. This allowed the model to make more precise determinations regarding organ and lesion locations. ASPP facilitated the acquisition of multi-scale features with diverse receptive fields, enabling the network to capture a richer array of semantic information. Finally, Conv_Unit2 was utilized to further fine-tune the multi-scale features generated by ASPP.

3.4. Double-Flow Linear Pooling Enhancement Unit

U-Net utilizes jump connections to combine shallow and deep semantic features. Nonetheless, the straightforward fusion method is susceptible to generating semantic gaps due to feature disparities. In order to tackle this issue, we optimized the jump connections and introduced the double-flow linear pooling enhancement unit (DLE). As illustrated in Figure 5, the DLE unit employed double-flow paths to establish cross-channel dependencies and gather a broader range of contextual information.

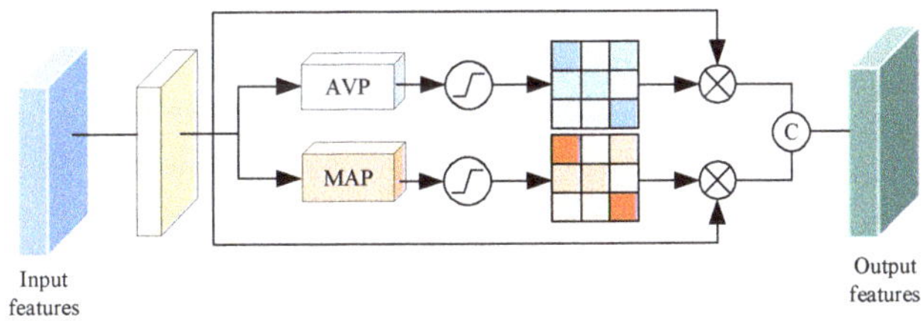

Figure 5. The structure of the double-flow linear pooling enhancement unit (DLE).

For the input feature map $F_{in} \in \mathbb{R}^{H\times W\times C}$, we applied deep convolution with a 3 × 3 convolution kernel and an expansion ratio of 2 to process the input feature map, resulting in a new feature map $F_{in}' \in \mathbb{R}^{H\times W\times}$. This operation, as opposed to standard convolution, captured feature map information across a larger range of sensory fields without introducing any additional parameters.

$$F_{in}' = DwConv_{k=3}^{rate=2}(F_{in}) \tag{4}$$

where $DwConv_{k=3}^{rate=2}$ denotes deep convolution with a kernel size of 3 × 3 and an expansion ratio = 2.

Subsequently, we conducted max pooling and average pooling on F_{in}' to extract more comprehensive channel information and generate feature maps $F_{ap} \in \mathbb{R}^{1\times 1\times C}$ and $F_{mp} \in \mathbb{R}^{1\times 1\times C}$, respectively.

$$\begin{aligned} F_{ap} &= Avgpool(F_{in}') \\ F_{mp} &= Maxpool(F_{in}') \end{aligned} \tag{5}$$

Finally, we employed the softmax function to normalize the weights of both F_{ap} and F_{mp} along the channel dimension. The outcome is two new attention views obtained by multiplying these weight matrices with F_{in}'. These two views are concatenated along the channel dimension, resulting in a concatenated feature map with dimensions H × W × 2C. Following this, dimensionality reduction is executed using the linear function W_{μ}, and the outcome is input into the decoder. The precise methodology is as follows:

$$F_{out} = W_{\mu}(Concatenate(F_{in}' \times \sigma(F_{ap}; F_{mp}))) \tag{6}$$

where $\sigma(\cdot)$ signifies the *sigmoid* function, F_{out} represents the output feature map resulting from channel-wise concatenation, and the linear function W_{μ} is implemented through a 1×1 convolution operation. This convolution operation serves the purpose of reducing the channel dimensions of the feature map, which aids in the subsequent feature fusion process.

The double-flow linear pooling enhancement unit integrates shallow and deep features in a "progressive" manner. It simultaneously feeds the generated contextual information and the original encoder features into the decoder. In addition to diminishing the semantic information gap between different pathways, the DLE unit strengthens information exchange between the encoder and decoder pathways, leading to enhanced model stability. Moreover, the features extracted by this unit have a beneficial impact on the localization of target regions. Furthermore, the deep convolution and pooling operations partially reduce both the parameter count and computational load.

3.5. Cross-Level Information Interaction

While extracting detailed features of the liver and tumor, shallow a priori knowledge like organ boundaries and texture is often neglected. To address this concern, we introduced a cross-level information interaction mechanism based on bidirectional residual connections. This mechanism enhanced the network's capability to learn and represent features by modeling both the encoder and decoder. As depicted by the blue arrows in Figure 6, the cross-level information interaction mechanism comprised shallow forward residuals and deep reverse residuals, detailed below.

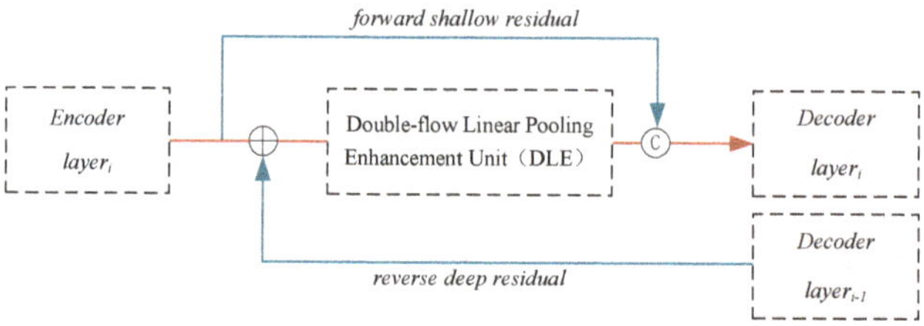

Figure 6. The structure of the cross-level information interaction mechanism (CII).

Suppose $x_i \in R^{H \times W \times C}$ denotes the shallow forward residual input originating from layer i of the encoder, and $f_{i-1} \in R^{\frac{H}{2} \times \frac{W}{2} \times \frac{C}{2}}$ represents the deep reverse residual input received from layer $i-1$ of the decoder.

First, an upsampling operation is conducted on f_{i-1} to bring its resolution in line with that of x_i for subsequent operations. This upsampling is achieved through bilinear interpolation.

$$f_{i-1}' = upsample(f_{i-1}) \tag{7}$$

Then, x_i and f_{i-1}' are summed element-wise and directed into the *DLE* unit for feature extraction. The features obtained are combined with x_i and subsequently reduced in dimensionality using linear mapping to produce the ultimate output y_b.

$$y_b = (Concatenate(DLE(x_i + f_{i-1}'), x_i))G \tag{8}$$

where *DLE* denotes the double-flow linear pooling enhancement unit, and *G* denotes the linear mapping function, which is implemented by 1 × 1 convolution.

The cross-level information interaction mechanism, founded on bidirectional residuals, achieves the fusion of contextual information across various layers. It effectively addresses the problem of forgetting a priori knowledge during training and expedites feature fusion within the network, serving as an automatic learning mechanism.

4. Results

4.1. Implementation Details

MDAU-Net was implemented with the Tensorflow2.0 framework, and we used a Tesla V100 to accelerate the calculations. We employed the Adam optimizer during the training process, which is widely selected in medical image segmentation tasks. Considering the computing resources, we set the batch size to eight. The initial learning rate was set to 1×10^{-4}, and when the loss did not decrease after two epochs, we updated the next learning rate to one-tenth of the current one. All experiments and models were trained using the same parameters.

4.1.1. Dataset

The segmentation datasets used in this paper were Liver Tumour Segmentation (LiTS) and Segmentation of the Liver Competition 2007 (SLiver07).

LiTS is the public dataset of the MICCAI 2017 Liver Tumor Segmentation Challenge, which contains 131 training sets and 70 test sets. Both of them contain patients' contrast-enhanced 3D abdominal CT scans with a resolution of 512 × 512. The in-plane resolution is 0.55~1.0 mm. The training dataset was labeled by experienced clinicians, but the testing dataset was not. Nevertheless, due to the large scale of the dataset and the high quality of the CT scans, it is currently a wildly used dataset in liver and tumor segmentation tasks.

SLiver07 is an earlier dataset that originated from the Segmentation of the Liver Competition 2007 (SLIVER07). It contains 20 training sets and 10 testing sets, which comprise clinical CT scans. The size of the images is 512 × 512, with an in-plane resolution of 0.56~0.8 mm. The training sets are labeled, while the 10 testing sets of CT scans are not, and both sets only contain liver information.

Since the testing sets of the two datasets were unlabeled, we only used the training set for all experiments. In detail, we used these two datasets for liver segmentation experiments, though only LiTS was selected to conduct tumor segmentation, as Sliver07 does not contain tumor information.

4.1.2. Data Preprocessing and Enhancement

For both LiTS and SLiver07, the training sets were further randomly partitioned into training and test subsets in an 8:2 ratio. This division was instrumental in evaluating the model's performance and generalization capacity. During the experimental data preparation phase, all original CT images underwent adjustment so that the Hounsfield values (HU values) fell within the range of [−200, 200]. This ensured that the images retained maximum liver volume while mitigating the noise interference stemming from other organs and background factors. Subsequently, the images were resampled, and their resolution was downsized from 512 × 512 to 256 × 256 to reduce computational overhead. Finally, normalization, slicing, and histogram equalization operations were performed sequentially.

Figure 7 shows some comparison images randomly selected from LiTS before and after preprocessing, where (1) to (3) are the original CT images without processing, and (4) to (6) are the images after a series of preprocessing operations. Obviously, the preprocessed images provided clearer boundary contours between abdominal organs, such as the liver. The contrast with the background was significantly enhanced, accompanied by more complete local details, which helped the network to capture more adequate feature information.

Compared to other semantic segmentation datasets, our dataset was small in size and originated from a small sample pool, so the data were enhanced by panning and rotating before the experiment to improve the diversity of the dataset.

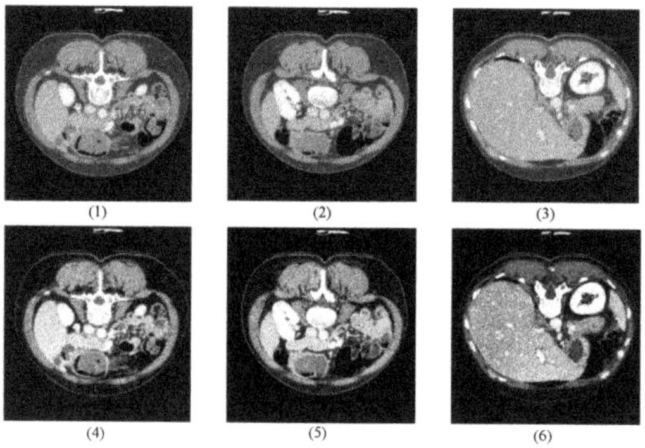

Figure 7. Comparison before and after data preprocessing.

4.1.3. Loss Function

During computer-aided diagnosis or clinical processes, achieving high recall is a critical performance indicator for models. The presence of unbalanced data in medical datasets makes a network easily fall into the local optimum. This, in turn, adversely impacts segmentation performance, often leading to a high precision but low recall. In order to balance the differences between categories among the training samples, Tversky loss was experimentally selected as the loss function to calculate the similarity between the predicted labels and the ground truth, with the following equation:

$$T(\alpha,\beta) = \frac{\sum_{i=1}^{n} p_{0i}g_{0i}}{\sum_{i=1}^{n} p_{0i}g_{0i} + \alpha\sum_{i=1}^{n} p_{0i}g_{1i} + \beta\sum_{i=1}^{n} p_{1i}g_{0i}} \quad (9)$$

where p_{0i} denotes the probability that the ith voxel is a tumor; p_{1i} denotes the probability that the ith voxel is not a tumor; $g_{0i} = 1$ denotes a lesion voxel; $g_{0i} = 0$ denotes a normal voxel; and g_{1i} the opposite. α and β are two hyperparameters, set to $\alpha + \beta = 1$, reducing the effect of positive and negative sample imbalance on model performance by adjusting the values of α and β. When $\alpha = \beta = 0.5$, Tversky loss [24] simplifies to the Dice coefficient while equating to the balanced F score (F1 score).

4.1.4. Evaluation Metrics

To evaluate the model performance and generalization ability more objectively and comprehensively, we selected five evaluation metrics for the experiments:

1. Dice coefficient (*Dice*)

$$Dice = \frac{2 \times |P \cap G|}{|P| + |G|} \quad (10)$$

2. Precision

$$Pre = \frac{TP}{TP + FP} \quad (11)$$

3. Recall

$$Recall = \frac{TP}{TP + FN} \quad (12)$$

4. Volumetric overlap error (*VOE*)

$$VOE = 1 - \frac{P \cap G}{P \cup G} \quad (13)$$

5. Relative volume error (*RVD*)

$$RVD = \frac{|P| - |G|}{|G|} \tag{14}$$

where *TP* is true positive, indicating that the liver region was correctly segmented; *TN* is true negative, which indicates that other organ regions were correctly segmented as the background; *FP* is false positive, which means that other organ regions were incorrectly segmented as the liver; *FN* is false negative, implying that the liver regions were incorrectly segmented as the background; *P* indicates the target pixel of the predicted label; and *G* represents the target pixel of the ground truth.

4.2. Loss Function Comparison Experiment

The imbalanced distribution of target and background poses a significant challenge in the domain of liver and liver tumor segmentation. This imbalance not only diminishes models' accuracy and generalization but also tends to favor high precision at the expense of low recall. To address this issue algorithmically, experiments were conducted to refocus the model on segmenting challenging samples by rebalancing the class distribution. We evaluated multiple common binary loss functions in the segmentation field, including Tversky loss, binary cross-entropy loss (BCE loss), Dice loss [25], and focal loss [26], on the LiTS dataset. The goal was to identify a loss function that was well-suited to our segmentation task and illustrate how it could alleviate the impact of imbalanced sample distribution on model performance, showcasing its superiority in enhancing model effectiveness compared to other loss functions.

The results are presented in Table 1. When the model employed Tversky loss, it achieved the best performance in Dice, Recall, and VOE, with scores of 0.9433, 0.9451, and 0.1053, respectively. In the case of BCE loss, the RVD exhibited the most favorable effect at 0.0189. However, when focal loss was utilized, the model's accuracy reached its highest point at 0.9662, but this came at the cost of a noticeable trade-off between precision and recall, resulting in a pronounced impact on class distribution. In comparison to the other three sets of loss functions, Tversky loss stood out with the most substantial optimization effect on model performance and a superior ability to balance positive and negative samples within the dataset. The gap between precision and recall steadily narrowed as both metrics improved, so this was selected as the experimental loss function.

Table 1. Loss function comparison test on LiTS.

Loss	Dice	Precision	Recall	VOE	RVD
Dice loss	0.9420	0.9490	0.9393	0.1076	0.0205
Focal loss	0.9044	**0.9662**	0.9116	0.1745	0.1872
Tversky loss	**0.9433**	0.9515	**0.9451**	**0.1053**	0.0383
BCE loss	0.9328	0.9486	0.9396	0.1239	**0.0189**

Bold text in the table represents the optimal results.

4.3. Validity Experiment of Cross-Level Information Interaction

In MDAU-Net, the cross-level information interaction mechanism, based on bidirectional residual connections, is frequently utilized in conjunction with the double-flow linear pooling enhancement unit. To demonstrate its effectiveness, we used U-Net with DLE as the baseline and assessed the segmentation performance by sequentially introducing residual pathways. We categorized the experiments into four groups. The first group served as the baseline, while the second and third groups were comparative experiments involving the addition of reverse and forward residual connections, respectively. The fourth group combined both forward and reverse residual connections. Table 2 displays the segmentation results for each group. Notably, performance was weakest when no residual connections were added. However, the introduction of either forward or reverse residuals led to varying degrees of performance improvement, with forward residuals demonstrating a more substantial positive impact on the network than reverse residuals. When both sets of residual connections were simultaneously incorporated, all evaluation metrics surpassed those of the first three groups. Compared to the baseline experiments, improvements of 0.0137, 0.0032, 0.0389, 0.0432, and 0.0313 were observed. Figure 8a presents the radar chart for this experiment, where the addition of two sets of residuals resulted in the largest coverage area on the coordinate axes, confirming that the cross-level information interaction mechanism based on bidirectional residual connections effectively mitigated knowledge forgetting issues and enhanced the network's learning capabilities.

Table 2. The performance of validity experiments on the LiTS dataset.

Method	Dice	Precision	Recall	VOE	RVD
Baseline	0.9067	0.9392	0.9019	0.1694	0.0759
Baseline + reverse residual	0.9080	0.9399	0.9054	0.1674	0.0872
Baseline + forward residual	0.9145	0.9403	0.9366	0.1398	0.0478
Baseline + bidirectional residual	**0.9204**	**0.9424**	**0.9408**	**0.1262**	**0.0446**

Bold text in the table represents the optimal results.

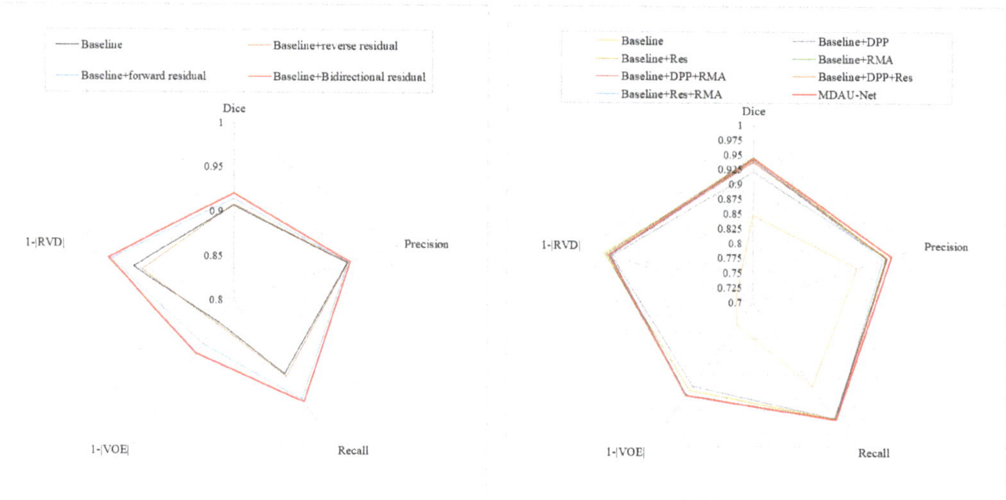

(a) (b)

Figure 8. This image shows the radar chart results from the validity experiment in Section 4.3 and the ablation experiment in Section 4.4: (**a**) radar chart of validity experiments, (**b**) radar chart of ablation experiments.

4.4. Ablation Results

To evaluate the effectiveness of various modules, we designed eight ablation experiments using the LiTS dataset. We chose U-Net with CII as the baseline to conduct the experiments. The first set was the baseline experiment. Sets 2 through 4 involved adding DLE, ResBlock, and CAE, respectively, on the basis of Experiment 1, which we used to verify the effect of each module on the baseline. Sets 5 to 7 added different combinations of modules onto the baseline to explore the dependencies among them. To verify the performance of the proposed method (MDAU-Net), set 8 added all modules to the first set to conduct training.

The results of the ablation experiments are displayed in Table 3 and Figure 8b. As depicted in Table 3, in the third set of experiments, the RVD attained a value of 0.0293, demonstrating that the inclusion of the residual encoder effectively preserved shallow features, thereby enhancing accuracy in organ boundary contour segmentation. In the fourth set of experiments, the Dice coefficient reached a value of 0.9447, indicating that valuable information was efficiently multiplexed within the cascaded adaptive feature extraction unit (CAE), resulting in increased similarity between the segmentation results and the ground truth. The results from other experimental sets showed that the addition of the ResBlock, CAE module, and DLE module each had a distinct positive impact on performance. Additionally, the radar plot in Figure 8b illustrates that MDAU-Net (red contour) achieved comparable Dice and RVD values while exhibiting superior accuracy, recall, and reduced error between predictions and ground truth.

Table 3. The performance of ablation experiments on LiTS.

Method	Dice	Precision	Recall	VOE	RVD
Baseline	0.8481	0.8879	0.8745	0.2536	0.2698
Baseline + DLE	0.9204	0.9424	0.9408	0.1262	0.0446
Baseline + ResBlock	0.9375	0.9409	0.9437	0.1161	**0.0293**
Baseline + CAE	**0.9447**	0.9422	0.9445	0.1064	0.0339
Baseline + DLE + CAE	0.9371	0.9437	0.9436	0.1062	0.0412
Baseline + DAE + ResBlock	0.9407	0.9425	0.9431	0.1056	0.0395
Baseline + ResBlock + CAE	0.9419	0.9354	0.9443	0.1070	0.0407
MDAU-Net	0.9433	**0.9515**	**0.9451**	**0.1053**	0.0383

Bold text in the table represents the optimal results.

5. Discussion

5.1. Quantitative Analysis of Liver Segmentation

To verify the effectiveness of MDAU-Net, we tested the method on the LiTS and SLiver07 datasets and compared it with other widely used segmentation methods.

Quantitative Analysis of Liver Segmentation on LiTS. The results of the liver segmentation on the LiTS dataset are shown in Table 4. The Dice, Precision, Recall, VOE, and RVD of MDAU-Net were 0.9433, 0.9515, 0.9451, 0.1053, and 0.0383, which were increased by 0.0952, 0.0636, 0.0706, 0.1483, and 0.2159, respectively, compared with the baseline U-Net values. Meanwhile, MDAU-Net had a significantly better balance between accuracy and recall, and the performance was outstanding in liver organ segmentation when compared to previous networks. Figure 9a shows the radar plots of the quantitative analysis of different models using LiTS. The red line represents MDAU-Net, which has the largest area covered by metrics on the axes, so that it can be more intuitively observed that its performance was better than the other comparison models.

Table 4. Liver semantic segmentation results of different models on LiTS.

Method	Dice	Precision	Recall	VOE	RVD
U-Net	0.8481	0.8879	0.8745	0.2536	0.2698
RU-Net [27]	0.8614	0.8902	0.8807	0.2415	0.2501
ResUNet [28]	0.9220	0.9263	0.9450	0.1427	0.0599
Attention U-net [29]	0.9197	0.9189	0.9236	0.1463	0.0575
UNet++ [10]	0.9106	0.9173	0.9075	0.1591	0.0818
SAR-U-Net [30]	0.9378	0.9504	0.9326	0.1142	0.0736
ResBCU-Net [31]	0.9359	0.9428	0.9302	0.1810	0.0587
RMS-UNet [32]	0.9171	0.9227	0.9157	0.1492	0.0646
MD-UNET [33]	0.9338	0.9433	0.9331	0.1224	0.0604
MDAU-Net (our model)	**0.9433**	**0.9515**	**0.9451**	**0.1053**	**0.0383**

Bold text in the table represents the optimal results.

Figure 10 displays the visualized results of the liver segmentation comparison test in this section. In these figures, the green lines represent the actual labels of the CT images, while the red lines indicate the prediction results. Additionally, we zoomed in on specific areas for easier observation. It can be inferred that due to the relatively simple structure of the jump connection between ResUNet and SAR-U-Net codecs, there was ineffective fusion of deep and shallow features, resulting in less precise image detail processing and a noticeable loss of edge information. Moreover, U-Net and UNet++ exhibited a limited utilization of a priori knowledge, such as shallow features, leading to difficulties in distinguishing between similar tissues and more prominent instances of mis-segmentation, where background organs were mistakenly segmented as the liver. In contrast, the visual segmentation results of MDAU-Net displayed the most complete segmentation and label curves, effectively fitting both continuous and truncated regions, with no significant instances of mis-segmentation or over-segmentation in detail processing.

To provide further insights into the test results of each model on the LiTS dataset and to assess the distinctions between the predictions of different models and the ground-truth labels, we utilized the confusion matrix. The results are presented in Figure 11, revealing that U-Net, U-Net++, and ResUNet exhibited difficulties in accurately recognizing the liver region, often misclassifying it as background. In contrast, MDAU-Net demonstrated a more balanced discrimination between the liver and background compared to other methods, with an extremely low probability of mis-segmentation and superior overall segmentation quality.

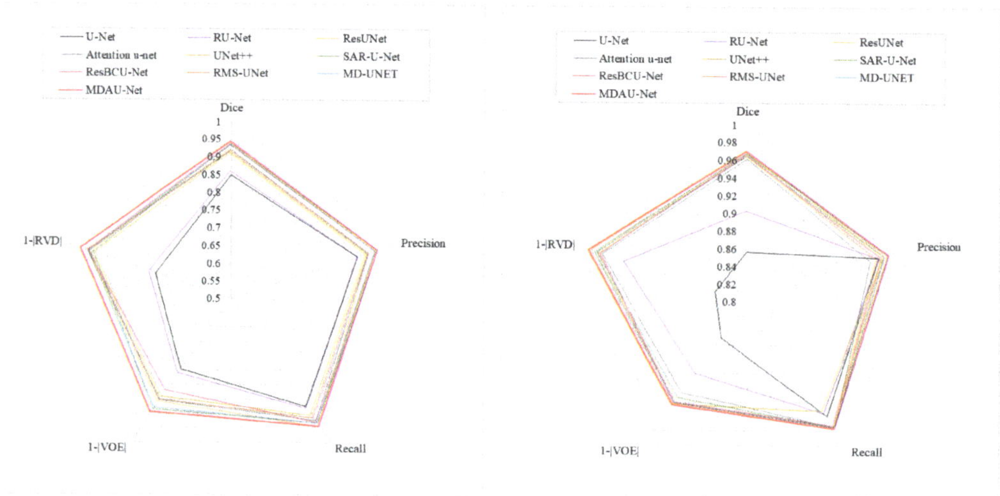

(a) (b)

Figure 9. This image shows the radar chart results from the liver segmentation in Section 5.1 on LiTS/Sliver07: (**a**) radar chart from LiTS, (**b**) radar chart from Sliver07.

Quantitative Analysis of Liver Segmentation on SLiver07. We opted to retrain MDAU-Net using the SLiver07 dataset to further assess its model performance. The experimental outcomes are detailed in Table 5, while Figure 9b presents corresponding radar plots of the experimental data. As indicated in the table, MDAU-Net achieved evaluation scores of 0.9706, 0.9743, 0.9757, 0.0569, and −0.0095 for various metrics. These scores represent improvements of 0.1138, 0.0137, 0.0169, 0.0932, and 0.1524 compared to the baseline U-Net, and they surpassed the performance of other methods to varying degrees. In the radar plot, MDAU-Net is depicted by a red outline, clearly demonstrating that it covers a wider area, indicative of its overall superiority compared to other examined methods.

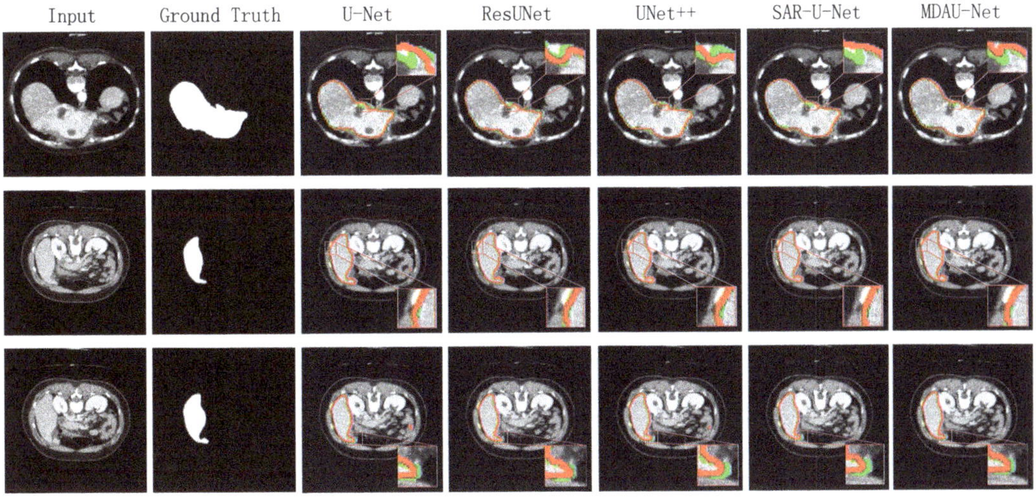

Figure 10. Visualization of results of liver segmentation using different methods on LiTS.

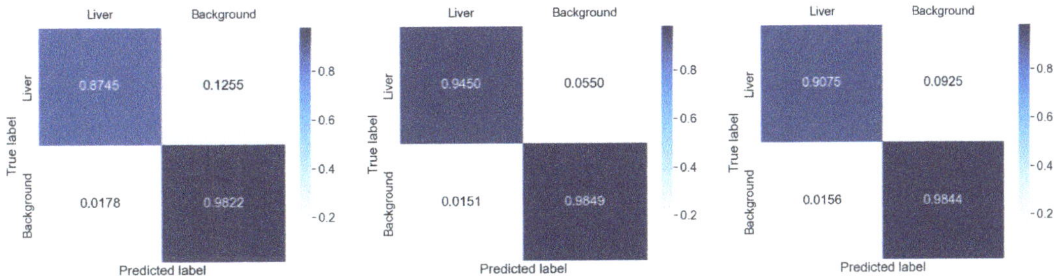

(a) U-Net (b) ResUNet (c) UNet++

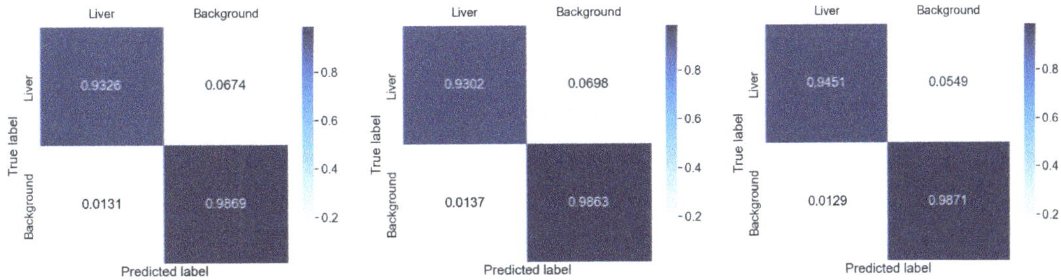

(d) SAR-U-Net (e) ResBCU-Net (f) MDAU-Net

Figure 11. Confusion matrix from liver segmentation in Section 5.1 on LiTS.

The visualized segmentation results for this set of experiments are presented in Figure 12. In these figures, the green lines represent the true labels, while the red lines depict the predicted results. To highlight the differences in segmentation outcomes, we magnified specific local areas. It is evident that U-Net and UNet++ exhibited more pronounced instances of mis-segmentation in the liver slices, with significant disparities between the segmentation results and the real labels in other slices. While ResUNet and SAR-U-Net produced improved segmentation results in the liver region compared to the former two methods, they still missed some detailed information in challenging segmentation areas. Conversely, MDAU-Net demonstrated the most complete overlap between the segmentation curves and the real labels, while also processing details such as the liver contour edge more comprehensively. This resulted in improved segmentation outcomes for liver slices of varying shapes and sizes compared to other methods.

Table 5. Liver semantic segmentation results for different models on SLiver07.

Method	Dice	Precision	Recall	VOE	RVD
U-Net	0.8568	0.9606	0.9588	0.1501	0.1619
RU-Net [27]	0.9032	0.9617	0.9546	0.1012	0.0523
ResUNet [28]	0.9697	0.9693	0.9740	0.0591	0.0184
Attention U-net [29]	0.9617	0.9501	0.9749	0.0733	−0.0254
UNet++ [10]	0.9703	0.9696	0.9515	0.0574	−0.0117
SAR-U-Net [30]	0.9655	0.9672	0.9746	0.0664	−0.0184
ResBCU-Net [31]	0.9658	0.9647	0.9723	0.0610	−0.0229
RMS-UNet [32]	0.9673	0.9601	0.9755	0.0591	−0.0238
MD-UNET [33]	0.9679	0.9732	0.9746	0.0601	−0.0162
MDAU-Net (our model)	**0.9706**	**0.9743**	**0.9757**	**0.0569**	**−0.0095**

Bold text in the table represents the optimal results.

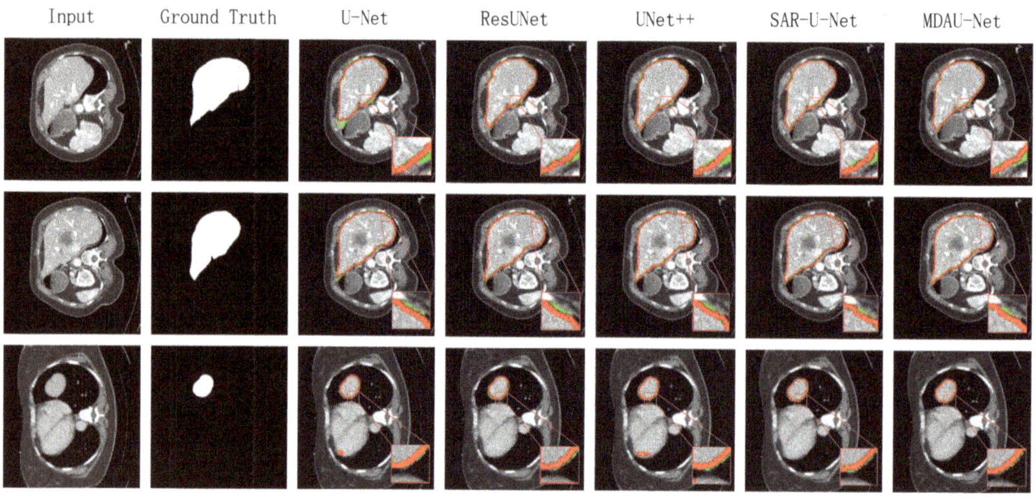

Figure 12. Visualization of results of liver segmentation using different methods on SLiver07.

Figure 13 presents the confusion matrix illustrating the segmentation results of each model on the SLiver07 dataset. It is evident that MDAU-Net exhibited a more balanced segmentation ability for both the liver and background regions, achieving superior segmentation results compared to other methods.

5.2. Quantitative Analysis of Liver Tumor Segmentation

On the LiTS dataset, we conducted a further comparison of MDAU-Net's performance in tumor segmentation tasks with other methods, and the results are presented in Table 6. When combined with the radar plot depicted in Figure 14, it is evident that MDAU-Net outperformed other methods in terms of Dice, VOE, and RVD, achieving values of 0.8387, 0.2699, and −0.0743, respectively. These values were 0.213, 0.1898, and 0.1929 higher than those obtained with UNet, indicating an overall superior segmentation performance compared to the other methods.

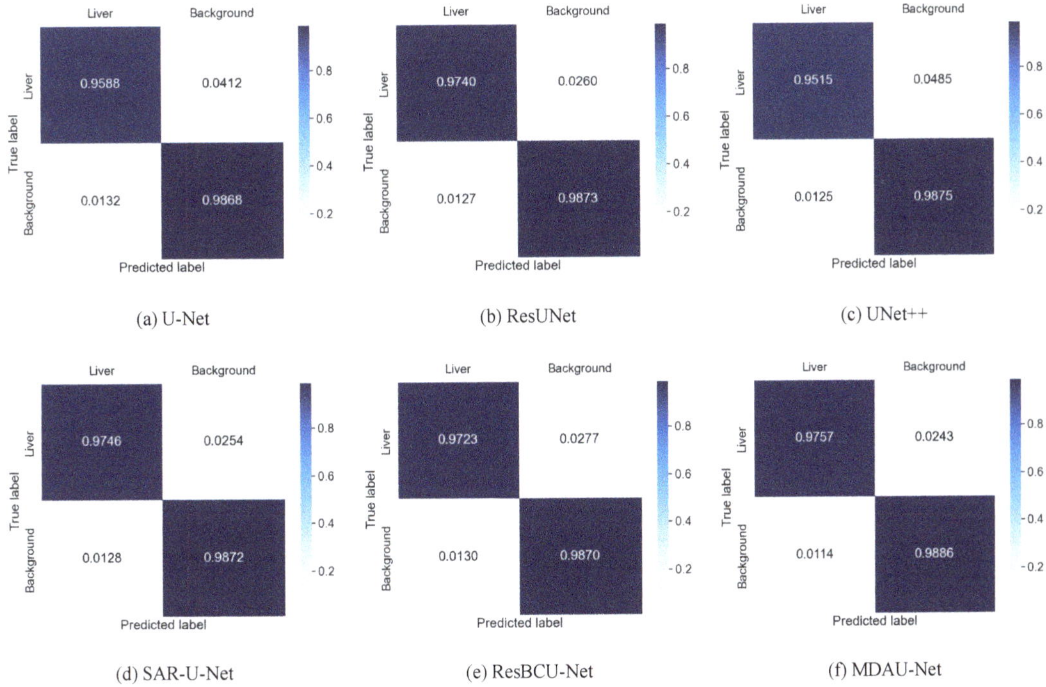

Figure 13. Confusion matrix from liver segmentation in Section 5.1 on SLiver07.

Table 6. Tumor segmentation results of different models on **LiTS**.

Method	Dice	Precision	Recall	VOE	RVD
U-Net	0.6257	0.6013	0.6128	0.4597	−0.2672
RU-Net [27]	0.6528	0.6233	0.6657	0.3926	−0.2519
ResUNet [28]	0.8254	0.8027	0.8550	0.2874	−0.0798
Attention U-net [29]	0.6683	0.6620	0.6807	0.3819	−0.0818
UNet++ [10]	0.7397	0.9340	0.7599	0.3995	−0.1930
SAR-U-Net [30]	0.8096	**0.8317**	0.8101	0.3495	−0.0770
ResBCU-Net [31]	0.6818	0.6243	0.7935	0.4588	−0.2278
RMS-UNet [32]	0.6712	0.6258	0.7829	0.4031	−0.2517
MD-UNET [33]	0.7838	0.7289	0.8593	0.3447	−0.1596
MDAU-Net (our model)	**0.8387**	0.8211	**0.8736**	**0.2699**	**−0.0743**

Bold text in the table represents the optimal results.

The visualization of the tumor segmentation results is presented in Figure 15. It is apparent that UNet and UNet++ exhibited insufficient segmentation and diagnostic errors when dealing with lesions characterized by blurred boundaries and small sizes. On the other hand, ResUNet and SAR-U-Net faced challenges in distinguishing between similar tissues, leading to suboptimal segmentation results. In contrast, MDAU-Net excelled in effectively localizing lesion tissues and accurately segmenting border regions, particularly for non-contiguous and small-sized lesions, demonstrating significantly improved performance. This underscores the effectiveness of the proposed method in addressing the issue of useful information loss, reducing the semantic gap between different pathways and achieving segmentation results with clear boundaries between lesion regions and normal tissues. Consequently, the proposed method holds substantial clinical value.

The confusion matrix illustrating the results of liver tumor segmentation on the LiTS dataset is displayed in Figure 16. Overall, all of these models demonstrated a high level of segmentation accuracy for non-diseased regions. In contrast, the segmentation results obtained by MDAU-Net were notably superior, with only a very small number of samples misclassified as non-diseased regions. Consequently, the likelihood of false-negative segmentation results is minimal, leading to more balanced segmentation outcomes.

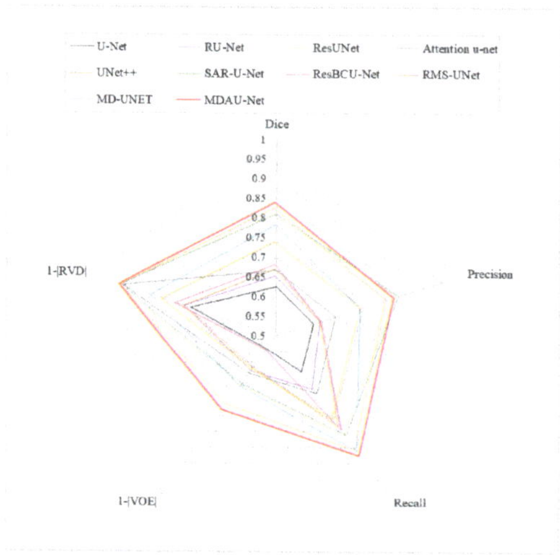

Figure 14. This image shows the radar chart results from liver tumor segmentation in Section 5.2 on LiTS.

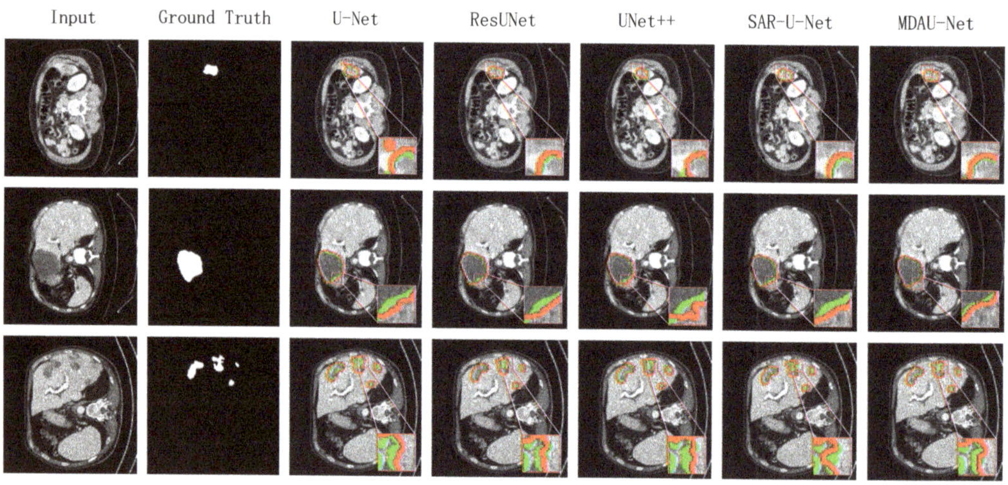

Figure 15. Visualization of results of liver tumor segmentation using different methods on LiTS.

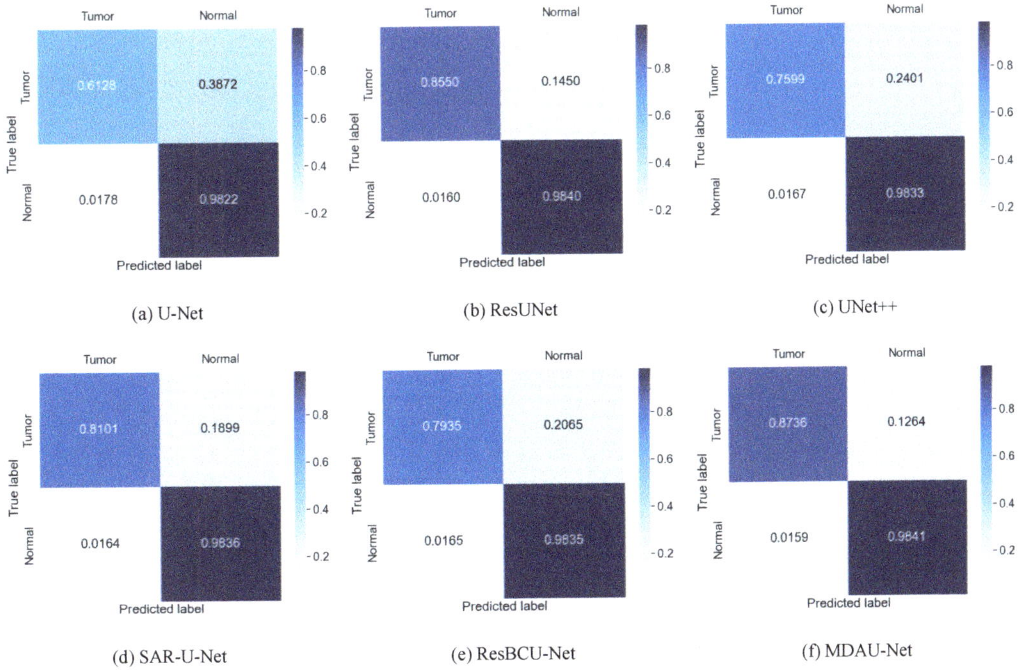

Figure 16. Confusion matrix from liver tumor segmentation in Section 5.2 on LiTS.

6. Conclusions

Owing to the exceptional achievements of U-Net in medical image processing, it has gained widespread adoption in liver and liver tumor segmentation tasks. Nonetheless, its straightforward network architecture hinders the comprehensive utilization of valuable features, leading to reduced feature mobility within the network. Moreover, the presence of a semantic gap impedes the effective fusion of shallow and deep features, consequently impacting the segmentation performance.

To address these issues, we proposed MDAU-Net, a novel segmentation network. MDAU-Net introduces a double-flow linear pooling enhancement unit within the jump connection segment, effectively narrowing the semantic divide and facilitating the fusion of shallow and deep features at each layer. Additionally, it incorporates a cascaded adaptive feature extraction unit as a bottleneck layer, which combines attention mechanisms with dense connectivity to enhance the network's capacity for exploring deep semantic information and improving feature mobility. Furthermore, a cross-level information interaction mechanism, based on bidirectional residuals, was introduced in the jump connection to mitigate the problem of a priori knowledge loss during training. Finally, we redesigned the encoder to incorporate the residual structure, not only enhancing the network's ability to retain and extract original features but also mitigating the gradient vanishing problem. Through experiments conducted on the LiTS and Sliver07 datasets, we confirmed that MDAU-Net consistently delivered outstanding performance across various datasets. It excelled not only in accurately segmenting the target region but also in handling intricate details such as edges with remarkable precision, demonstrating strong generalization capabilities.

Author Contributions: J.M., M.X., Z.M. and Z.J. designed the experiments. M.X. conducted the experiments, interpreted the data, and drafted the manuscript. J.M. and Z.M. provided professional suggestions. All authors have read and agreed to the published version of the manuscript.

Funding: This research was funded by the Natural Science Foundation of Ningxia (Nos. 2023AAC03264, 2022AAC03268); Basic Scientific Research in the Central Universities of North Minzu University (Nos. 2021KJCX09, FWNX21); Graduate Innovation Project of North Minzu University (YCX23148), and Image and Intelligent Information Processing Innovation Team of the National Ethnic Affairs Commission.

Institutional Review Board Statement: Not applicable.

Informed Consent Statement: Not applicable.

Data Availability Statement: The datasets analyzed during the current study are available in the LiTS and Sliver07 repositories: https://competitions.codalab.org; www.sliver07.org (accessed on 13 June 2023).

Conflicts of Interest: The authors declare no conflict of interest.

References

1. Hinton, G.E.; Osindero, S.; Teh, Y.W. A fast learning algorithm for deep belief nets. *Neural Comput.* **2006**, *18*, 1527–1554. [CrossRef] [PubMed]
2. Ronneberger, O.; Fischer, P.; Brox, T. U-net: Convolutional networks for biomedical image segmentation. In *Medical Image Computing and Computer-Assisted Intervention–MICCAI 2015: 18th International Conference, Munich, Germany, 5–9 October 2015, Proceedings, Part III 18*; Springer: Cham, Switzerland, 2015; pp. 234–241.
3. Dickson, J.; Lincely, A.; Nineta, A. A Dual Channel Multiscale Convolution U-Net Methodfor Liver Tumor Segmentation from Abdomen CT Images. In Proceedings of the 2022 International Conference on Sustainable Computing and Data Communication Systems (ICSCDS), Erode, India, 7–9 April 2022; pp. 1624–1628.
4. Sabir, M.W.; Khan, Z.; Saad, N.M.; Khan, D.M.; Al-Khasawneh, M.A.; Perveen, K.; Qayyum, A.; Azhar Ali, S.S. Segmentation of Liver Tumor in CT Scan Using ResU-Net. *Appl. Sci.* **2022**, *12*, 8650.
5. Deng, Y.; Hou, Y.; Yan, J.; Zeng, D. ELU-net: An efficient and lightweight U-net for medical image segmentation. *IEEE Access* **2022**, *10*, 35932–35941.
6. Seong, W.; Kim, J.H.; Kim, E.J.; Park, J.W. Segmentation of abnormal liver using adaptive threshold in abdominal CT images. In Proceedings of the IEEE Nuclear Science Symposuim & Medical Imaging Conference, Knoxville, TN, USA, 30 October–6 November 2010; pp. 2372–2375.
7. Chen, Y.; Wang, Z.; Zhao, W.; Yang, X. Liver segmentation from CT images based on region growing method. In Proceedings of the 2009 3rd International Conference on Bioinformatics and Biomedical Engineering, Beijing, China, 11–13 June 2009; pp. 1–4.
8. Gambino, O.; Vitabile, S.; Re, G.L.; La Tona, G.; Librizzi, S.; Pirrone, R.; Ardizzone, E.; Midiri, M. Automatic volumetric liver segmentation using texture based region growing. In Proceedings of the 2010 International Conference on Complex, Intelligent and Software Intensive Systems, Krakow, Poland, 15–18 February 2010; pp. 146–152.
9. Okada, T.; Shimada, R.; Hori, M.; Nakamoto, M.; Chen, Y.W.; Nakamura, H.; Sato, Y. Automated segmentation of the liver from 3D CT images using probabilistic atlas and multilevel statistical shape model. *Acad. Radiol.* **2008**, *15*, 1390–1403. [CrossRef] [PubMed]
10. Zhou, Z.; Rahman Siddiquee, M.M.; Tajbakhsh, N.; Liang, J. Unet++: A nested u-net architecture for medical image segmentation. In *Deep Learning in Medical Image Analysis and Multimodal Learning for Clinical Decision Support: 4th International Workshop, DLMIA 2018, and 8th International Workshop, ML-CDS 2018, Held in Conjunction with MICCAI 2018, Granada, Spain, 20 September 2018, Proceedings 4*; Springer: Cham, Switzerland, 2018; pp. 3–11.
11. Huang, H.; Lin, L.; Tong, R.; Hu, H.; Zhang, Q.; Iwamoto, Y.; Han, X.; Chen, Y.W.; Wu, J. Unet 3+: A full-scale connected unet for medical image segmentation. In Proceedings of the ICASSP 2020-2020 IEEE International Conference on Acoustics, Speech and Signal Processing (ICASSP), Barcelona, Spain, 4–8 May 2020; pp. 1055–1059.
12. Bi, R.; Ji, C.; Yang, Z.; Qiao, M.; Lv, P.; Wang, H. Residual based attention-Unet combing DAC and RMP modules for automatic liver tumor segmentation in CT. *Math. Biosci. Eng.* **2022**, *19*, 4703–4718. [PubMed]
13. Kushnure, D.T.; Talbar, S.N. HFRU-Net: High-level feature fusion and recalibration unet for automatic liver and tumor segmentation in CT images. *Comput. Methods Programs Biomed.* **2022**, *213*, 106501.
14. Zhou, Y.; Kong, Q.; Zhu, Y.; Su, Z. MCFA-UNet: Multiscale cascaded feature attention U-Net for liver segmentation. *IRBM* **2023**, *44* , 100789. [CrossRef]
15. Chen, L.C.; Zhu, Y.; Papandreou, G.; Schroff, F.; Adam, H. Encoder-decoder with atrous separable convolution for semantic image segmentation. In Proceedings of the European Conference on Computer Vision (ECCV), Munich, Germany, 8–14 September 2018; pp. 801–818.
16. Meng, T.; Ghiasi, G.; Mahjorian, R.; Le, Q.V.; Tan, M. Revisiting Multi-Scale Feature Fusion for Semantic Segmentation. *arXiv* **2022**, arXiv:2203.12683.
17. Zhang, D.; Zhang, H.; Tang, J.; Wang, M.; Hua, X.; Sun, Q. Feature pyramid transformer. In *Computer Vision–ECCV 2020: 16th European Conference, Glasgow, UK, 23–28 August 2020, Proceedings, Part XXVIII 16*; Springer: Cham, Switzerland, 2020; pp. 323–339.

18. Chen, L.C.; Papandreou, G.; Schroff, F.; Adam, H. Rethinking atrous convolution for semantic image segmentation. *arXiv* **2017**, arXiv:1706.05587.
19. Hu, J.; Shen, L.; Sun, G. Squeeze-and-excitation networks. In Proceedings of the IEEE Conference on Computer Vision and Pattern Recognition, Salt Lake City, UT, USA, 18–22 June 2018; pp. 7132–7141.
20. Woo, S.; Park, J.; Lee, J.Y.; Kweon, I.S. Cbam: Convolutional block attention module. In Proceedings of the European Conference on Computer Vision (ECCV), Munich, Germany, 8–14 September 2018; pp. 3–19.
21. Vaswani, A.; Shazeer, N.; Parmar, N.; Uszkoreit, J.; Jones, L.; Gomez, A.N.; Kaiser, Ł.; Polosukhin, I. Attention is all you need. *arXiv* **2017**, arXiv:1706.03762.
22. He, K.; Zhang, X.; Ren, S.; Sun, J. Deep residual learning for image recognition. In Proceedings of the IEEE Conference on Computer Vision and Pattern Recognition, Las Vegas, NV, USA, 27–30 June 2016; pp. 770–778.
23. Huang, G.; Liu, Z.; Van Der Maaten, L.; Weinberger, K.Q. Densely connected convolutional networks. In Proceedings of the IEEE Conference on Computer Vision and Pattern Recognition, Honolulu, HI, USA, 21–26 July 2017; pp. 4700–4708.
24. Salehi, S.S.M.; Erdogmus, D.; Gholipour, A. Tversky loss function for image segmentation using 3D fully convolutional deep networks. In *Machine Learning in Medical Imaging: 8th International Workshop, MLMI 2017, Held in Conjunction with MICCAI 2017, Quebec City, QC, Canada, 10 September 2017, Proceedings 8*; Springer: Cham, Switzerland, 2017; pp. 379–387.
25. Milletari, F.; Navab, N.; Ahmadi, S.A. V-net: Fully convolutional neural networks for volumetric medical image segmentation. In Proceedings of the 2016 fourth international conference on 3D vision (3DV), Stanford, CA, USA, 25–28 October 2016; pp. 565–571.
26. Lin, T.Y.; Goyal, P.; Girshick, R.; He, K.; Dollár, P. Focal loss for dense object detection. In Proceedings of theIEEE International Conference on Computer Vision, Venice, Italy, 22–29 October 2017; pp. 2980–2988.
27. Alom, M.Z.; Hasan, M.; Yakopcic, C.; Taha, T.M.; Asari, V.K. Recurrent residual convolutional neural network based on u-net (r2u-net) for medical image segmentation. *arXiv* **2018**, arXiv:1802.06955.
28. Xiao, X.; Lian, S.; Luo, Z.; Li, S. Weighted res-unet for high-quality retina vessel segmentation. In Proceedings of the 2018 9th International Conference on Information Technology in Medicine and Education (ITME), Hangzhou, China, 19–21 October 2018; pp. 327–331.
29. Oktay, O.; Schlemper, J.; Folgoc, L.L.; Lee, M.; Heinrich, M.; Misawa, K.; Mori, K.; McDonagh, S.; Hammerla, N.Y.; Kainz, B.; et al. Attention u-net: Learning where to look for the pancreas. *arXiv* **2018**, arXiv:1804.03999.
30. Wang, J.; Lv, P.; Wang, H.; Shi, C. SAR-U-Net: Squeeze-and-excitation block and atrous spatial pyramid pooling based residual U-Net for automatic liver segmentation in Computed Tomography. *Comput. Methods Programs Biomed.* **2021**, *208*, 106268. [CrossRef]
31. Badshah, N.; Ahmad, A. ResBCU-net: Deep learning approach for segmentation of skin images. *Biomed. Signal Process. Control* **2022**, *71*, 103137. [CrossRef]
32. Khan, R.A.; Luo, Y.; Wu, F.X. RMS-UNet: Residual multi-scale UNet for liver and lesion segmentation. *Artif. Intell. Med.* **2022**, *124*, 102231. [CrossRef]
33. Ge, R.; Cai, H.; Yuan, X.; Qin, F.; Huang, Y.; Wang, P.; Lyu, L. MD-UNET: Multi-input dilated U-shape neural network for segmentation of bladder cancer. *Comput. Biol. Chem.* **2021**, *93*, 107510. [CrossRef]

applied sciences

MDPI

Article

A Fuzzy Consensus Clustering Algorithm for MRI Brain Tissue Segmentation

S. V. Aruna Kumar [1,*], Ehsan Yaghoubi [2] and Hugo Proença [3,*]

1 Department of Computer Science and Engineering, Malnad College of Engineering, Hassan 573202, Karnataka, India
2 Department of Informatics, University of Hamburg, 22527 Hamburg, Germany; ehsan.yaghoubi@uni-hamburg.de
3 Department of Computer Science, University of Beira Interior, 6201-001 Covilhã, Portugal
* Correspondence: arunkumarsv55@gmail.com (S.V.A.K.); hugomcp@di.ubi.pt (H.P.)

Abstract: Brain tissue segmentation is an important component of the clinical diagnosis of brain diseases using multi-modal magnetic resonance imaging (MR). Brain tissue segmentation has been developed by many unsupervised methods in the literature. The most commonly used unsupervised methods are K-Means, Expectation-Maximization, and Fuzzy Clustering. Fuzzy clustering methods offer considerable benefits compared with the aforementioned methods as they are capable of handling brain images that are complex, largely uncertain, and imprecise. However, this approach suffers from the intrinsic noise and intensity inhomogeneity (IIH) in the data resulting from the acquisition process. To resolve these issues, we propose a fuzzy consensus clustering algorithm that defines a membership function resulting from a voting schema to cluster the pixels. In particular, we first pre-process the MRI data and employ several segmentation techniques based on traditional fuzzy sets and intuitionistic sets. Then, we adopted a voting schema to fuse the results of the applied clustering methods. Finally, to evaluate the proposed method, we used the well-known performance measures (boundary measure, overlap measure, and volume measure) on two publicly available datasets (OASIS and IBSR18). The experimental results show the superior performance of the proposed method in comparison with the recent state of the art. The performance of the proposed method is also presented using a real-world Autism Spectrum Disorder Detection problem with better accuracy compared to other existing methods.

Keywords: brain tissue segmentation; consensus clustering; segmentation; magnetic resonance image

Citation: Aruna Kumar, S.V.; Yaghoubi, E.; Proença, H. A Fuzzy Consensus Clustering Algorithm for MRI Brain Tissue Segmentation. *Appl. Sci.* **2022**, *12*, 7385. https://doi.org/10.3390/app12157385

Academic Editors: Ioannis A. Kakadiaris, Michalis Vrigkas and Christophoros Nikou

Received: 19 April 2022
Accepted: 20 July 2022
Published: 22 July 2022

1. Introduction

Segmenting brain tissue is the process of subdividing the image of the brain into major components such as Cerebrospinal Fluid (CSF), Gray Matter (GM), and White Matter (WM). The step of brain tissue segmentation is fundamental in diagnosing and monitoring a wide range of neurological diseases. Several researchers have strived to develop automatic brain tissue segmentation in the last two decades [1–4].

Brain tissue segmentation has been developed by many unsupervised methods in the literature. The most commonly used unsupervised methods are: K-Means [5–7], Expectation-Maximization [8], and Fuzzy Clustering [9,10]. Fuzzy clustering methods offer considerable benefits compared with the aforementioned methods as they are capable of handling brain images that are complex, largely uncertain, and imprecise.

Even thoug, traditional Fuzzy C-Means (FCM) showcases outstanding results on brain image segmentation, it has some limitation such as being sensitive to noise due to the use of the Euclidean distance metric and neighbourhood information ignorance. The FCM computes the distance between cluster center and voxels using a Euclidean distance measure. Euclidean distance is very sensitive to noise which results in the deterioration of segmentation results. In the literature, we found many variants of FCM methods that are developed to address the aforementioned shortcomings. To address the noise sensitivity, researchers added the spatial information into the FCM objective function [11,12]. The addition of a spatial function to an objective function helps to reduce the impact of noise and also helps to enhance performance. The spatial information may be local or global [13]. On the other hand, to address the limitations of Euclidean distance, many researchers developed a kernel version of FCM and named it Kernel FCM (KFCM) [14,15]. KFCM adopts kernel function as a distance measure. The kernel function transfers the input data to higher dimensional kernel space and makes the clustering task easier. The aforementioned FCM variant methods are based on a traditional fuzzy set. In a fuzzy set, the non-membership value is always the complement of the membership value. However, in real time, this assumption fails due to hesitation. The hesitation arises due to uncertainty in defining the membership function. To handle this hesitation, Atanassov [16] developed an advanced fuzzy set called Intuitionistic Fuzzy Set (IFS). In IFS, the non-membership value is computed using the fuzzy complement generator functions. In recent times, researchers have given more attention in developing IFS-based clustering methods [17–20]. Chaira [18] developed an Intuitionistic Fuzzy C-Means (IFCM) where the intuitionistic fuzzy entropy is added to the conventional FCM objective function. The intuitionistic fuzzy set handles the uncertainty which originates while defining a membership function by considering the hesitation degree. To handle noise and uncertainty during segmentation, Verma et al. [21] considered both the pixel and local neighborhood information. The main benefit of this method is that it is non-parametric.

Recently, researchers have come to realize that a single clustering method might fail to produce good results with complex data. Hence, they are concentrating on developing consensus clustering methods [22,23]. Consensus clustering is also known as cluster ensemble, and its main aim is to find a single partition of data with overlapping clusters. In the literature, it has been widely agreed that consensus clustering can generate robust results [24–27]. Motivated by the advantages of consensus clustering, in this paper we are proposing a brain tissue segmentation method based on consensus clustering. The proposed method consists of two steps: Pre-processing and Segmentation. In pre-processing, the brain images are pre-processed by employing registration, skull stripping, and bias field correction. In the segmentation step, initially the brain images are segmented using four different clustering methods. The two clustering methods are based on a traditional fuzzy set and the other two are based on an intuitionistic set. In the traditional fuzzy set category, Robust Spatial Kernel FCM (RSKFCM) [28] and Generalized Spatial Kernel FCM (GSKFCM) [29] are employed. On the other hand, in the intuitionistic fuzzy set category, two variants of Modified Intuitionistic Fuzzy C-Means [20] are employed. Furthermore, the results of four individual clustering methods are combined using a voting schema. The proposed approach is evaluated on two publicly available MRI datasets: OASIS and IBSR18 Dataset, and the results are compared using the results with state-of-art methods. The primary contributions of this paper are as follows:

- Proposed a consensus clustering method for MRI brain tissue segmentation.
- The results of four variants of fuzzy clustering methods are combined to achieve better results.
- To check the efficacy of the proposed method, we conducted experiments on two standard brain segmentation datasets.

The remainder of the paper is as follows: Section 2 presents the methodology of the proposed method. We then introduce the datasets and the evaluation metrics alongside the implementation details and discussions on the performance of the proposed method in Section 3. Finally, the conclusion of the paper is presented in Section 4.

2. Methodology

This section presets the methodology of the proposed method. The proposed consensus clustering method comprises two steps: Pre-procesing and Segmentation.

2.1. Pre-Processing

We perform three pre-processing steps, namely Registration, Bias Field correction, and Skull Stripping. Registration is the process of spatially aligning two or more images of the same content taken from a different view and/or at a different time, and alignment of the multi-modal image of the same patient is required. Bias Field refers to a low-frequency signal which corrupts the MRI images due to inhomogeneities in the magnetic field of the MRI machines. Bias field leads to intensity inhomogeneity, and in turn, it affects the segmentation accuracy. Hence, the bias field needs to be corrected before performing the segmentation. Skull Stripping is the process of removing non-brain tissues such as fat, skull, and neck. These non-brain tissues have an intensity that overlaps with the intensity of the other brain tissues. Thus, the brain tissues have to be extracted before the brain segmentation. There are many skull stripping methods such as Brain Extraction Tool (BET) [30], Brain Surface Extraction (BSE) [31], AFNI ("Analysis of Functional NeuroImages" (AFNI) software package publicly available at https://afni.nimh.nih.gov/ (accessed on 19 April 2022)), BridgeBurner [32], GCUT [33], and ROBEX [34]. Among all these methods, ROBEX provides significantly improved performance [34].

All the aforementioned steps are optional and depend on the image data used for the study. Hence, in this paper, different pre-processing steps are performed for different datasets. The pre-processed brain images are segmented using consensus clustering. The following subsection presents a detailed description regarding segmentation.

2.2. Segmentation

The proposed consensus clustering method consists of a combination of traditional fuzzy sets and intuitionistic sets to not only increase the robustness of the noise but also use the neighborhood information when forming the clusters. To do so, we use the Robust Spatial Kernel FCM (RSKFCM) [28] and Generalized Spatial Kernel FCM (GSKFCM) [29] methods alongside the two variants of the Modified Intuitionistic Fuzzy C-Means [20] technique. Finally, we fuse the results of the clustering methods using a voting schema. The next subsections explain the employed clustering methods and the voting schema in detail.

2.2.1. Robust Spatial Kernel FCM (RSKFCM)

Robust Spatial Kernel Fuzzy C-Means (RSKFCM) [28] is the variant of conventional Fuzzy C-Means (FCM). RSKFCM addresses the noise sensitivity and neighborhood information ignorance limitations of FCM. RSKFCM injects the neighborhood information into the FCM objective function and uses the Gaussian Kernel function instead of the Euclidean metric.

The main aim of the RSKFCM is to minimize the objective function shown in Equation (1)

$$J = \sum_{i=1}^{c} \sum_{j=1}^{n} w_{ij}^{m} \, \|\Phi(x_j) - \Phi(v_i)\|^2 \tag{1}$$

where c is the number of clusters, n is the number of voxels, m is a fuzzifier value, which controls the fuzziness of the resulting partition, w_{ij} is the RSKFCM membership degree of x_j in ith cluster, v_i is the ith cluster center, and Φ is an implicit nonlinear map which is computed as:

$$\|\Phi(x_j) - \Phi(v_i)\|^2 = K(x_j, x_j) + K(v_j, v_j) - 2K(x_j, v_i) \tag{2}$$

where K is the inner product of kernel function, i.e., $K(x,y) = \Phi(x)^T\Phi(y)$. In this paper, we have adopted the Gaussian kernel function which is defined as:

$$K(x,y) = \exp\left(-\|x-y\|^2 / \sigma^2\right) \tag{3}$$

In Gaussian kernel, $K(x,x) = 1$ and $K(v,v) = 1$, hence the kernel function becomes:

$$\|\Phi(x_j) - \Phi(v_i)\|^2 = 2(1 - K(x_j, v_i)) \tag{4}$$

Substituting Equation (4) in Equation (1), the objective function becomes:

$$J = 2\sum_{i=1}^{c}\sum_{j=1}^{n} w_{ij}^{m}\,(1 - K(x_j, v_i)) \tag{5}$$

The RSKFCM membership function w_{ij} is the combination of the kernel membership function u_{ij}, and the neighbourhood function s_{ij} and it is computed as.

$$w_{ij} = \frac{u_{ij}^{p} s_{ij}^{q}}{\sum_{k=1}^{c} u_{kj}^{p} s_{kj}^{q}} \tag{6}$$

where p and q are parameters to control the relative importance of the kernel membership and the neighbourhood membership functions.

The kernel and neighbourhood membership functions are computed using Equations (7) and (8)

$$u_{ij} = \frac{\left(1 - K(x_j, v_i)\right)^{-1/(m-1)}}{\sum_{k=1}^{c}\left(1 - K(x_j, v_k)\right)^{-1/(m-1)}}; \tag{7}$$

$$s_{ij} = \sum_{k \in N_k(x_j)} u_{ik} \tag{8}$$

where $N_k(x_j)$ represents neighbourhood voxels of x_j. This neighbourhood function represents the probability that the voxel x_j belongs to the ith cluster.

Similar to FCM, RSKFCM also works in an iterative process to update the membership and cluster center values. The cluster centers are updated using Equation (9)

$$v_i = \frac{\sum_{j=1}^{n} w_{ij}^{m} K(x_j, v_i) x_j}{\sum_{j=1}^{n} w_{ij}^{m} K(x_j, v_i)} \tag{9}$$

RSKFCM is an iterative process, and it stops when the stopping criteria is satisfied, i.e., the difference of successive iteration's objective function value is less than the user-specified stopping criteria value.

2.2.2. Generalized Spatial Kernel FCM (GSKFCM)

The generalized Spatial Kernel FCM (GSKFCM) [29] is another variant of the conventional FCM. Even though RSKFCM overcomes the limitations of the FCM, the performance is not good because it injects neighborhood information only into the objective function. However, the distance function plays a vital role in computing the membership value. Thus, the addition of neighborhood information can increase the performance. The RSKFCM also assumes all features have equal importance. However, in a real-world problem, all the features may not be equally important. GSKFCM overcomes these limitations by injecting the weighted neighbourhood information into the distance function and employing the Gaussian kernel as the distance metric.

The aim of the GSKFCM is to minimize the objective function shown in Equation (10).

$$J = 2\sum_{i=1}^{c}\sum_{j=1}^{n} z_{ij}^{m} d_{new}^{2}(x_j, v_i) \tag{10}$$

where z_{ij} is the GSKFCM membership function, and it is computed as:

$$z_{ij} = \frac{1}{\sum\limits_{k=1}^{c}\left(\frac{d_{new}^{2}(x_j,v_i)}{d_{new}^{2}(x_j,v_k)}\right)^{\frac{1}{(m-1)}}} \tag{11}$$

where d_{new} is the GSKFCM distance function which incorporates the neighbourhood function into the distance function, and it is computed as:

$$d_{new}^{2}(x_j, v_i) = d^2(x_j, v_i)\, f(p_{ij}) \tag{12}$$

where, $d^2(x_j, v_i)$ is the Gaussian Kernel distance function shown in Equation (4), and $f(p_{ij}) = \frac{1}{p_{ij}}$ is the neighbourhood function.

The GSKFCM considers the neighbourhood information and computes the membership value associated with each voxel as the weighted sum of the traditional FCM membership value and the membership value of the N_k neighbour points. The neighbourhood function (p_{ij}) is defined as:

$$p_{ij} = \sum_{k=0}^{N_k} h(x_j, x_k) g(u_{ik}) \tag{13}$$

where N_k is the number of neighbourhood voxels, $g(u_{ik}) = u_{ik}$ is the membership function (Equation (7)), $h(x_j, x_k)$ is the distance function which is computed as:

$$h(x_j, x_k) = \left(\sum_{l=0}^{N_k} \frac{d^2(x_j, x_k)}{d^2(x_j, x_l)}\right)^{-1} \tag{14}$$

Substituting Equation (14) in Equation (13), the neighbourhood function becomes:

$$p_{ij} = \sum_{k=0}^{N_k} g(u_{ik}) \left(\sum_{l=0}^{N_k} \frac{d^2(x_j, x_k)}{d^2(x_j, x_l)}\right)^{-1} \tag{15}$$

Substituting Equation (12) in Equation (11), the membership function z_{ij} becomes,

$$z_{ij} = \left(\sum_{k=1}^{c}\left(\frac{d^2(x_j, v_i)\, f(p_{ij})}{d^2(x_j, v_k)\, f(p_{jk})}\right)^{\frac{1}{(m-1)}}\right)^{-1} \tag{16}$$

$$= \frac{\left(\sum\limits_{k=1}^{c}\left(\frac{d^2(x_j,v_i)}{d^2(x_j,v_k)}\right)^{\frac{1}{m-1}}\right)^{-1} f^{\frac{1}{1-m}}(p_{ij})}{\sum\limits_{k=1}^{c}\left(\sum\limits_{l=1}^{c}\left(\frac{d^2(x_j,v_i)}{d^2(x_j,v_l)}\right)^{\frac{1}{m-1}}\right)^{-1} f^{\frac{1}{1-m}}(p_{jk})} \tag{17}$$

where $\left(\sum\limits_{k=1}^{c}\left(\frac{d^2(x_j,v_i)}{d^2(x_j,v_k)}\right)^{\frac{1}{m-1}}\right)^{-1} = u_{ij}$. Then the membership function z_{ij} becomes

$$z_{ij} = \frac{u_{ij} f^{\frac{1}{1-m}}(p_{ij})}{\sum\limits_{k=1}^{c} u_{jk} f^{\frac{1}{1-m}}(p_{jk})} \tag{18}$$

Similar to FCM and RSKFCM, GSKFCM operates as an iterative process by updating membership and cluster center value. The cluster centers are updated using Equation (19)

$$v_i = \frac{\sum\limits_{j=1}^{n} z_{ij}^m K(x_j, v_i) x_j}{\sum\limits_{j=1}^{n} z_{ij}^m K(x_j, v_i)} \tag{19}$$

GSKFCM decides the label based on the maximum membership value.

2.2.3. Modified Intuitionistic Fuzzy C-Means (MIFCM)

Modified Intuitionistic Fuzzy C-Means (MIFCM) [20] is the variant of the conventional Intuitionistic Fuzzy C-Means (IFCM) [18], and it is based on an intuitionistic fuzzy set. In MIFCM, the input data is clustered by optimizing the following objective function shown in Equation (20)

$$J = \sum_{j=1}^{n}\sum_{i=1}^{c} \beta_{ij}^m \, d_H(x_j, v_i) \tag{20}$$

where x_j represents jth voxel, v_i refers to ith cluster center, m refers to the fuzzification value, β_{ij} refers to the MIFCM membership value of jth voxel to ith cluster, and $d_H(x_j, v_i)$ is the modified Hausdorff distance between jth voxel to ith cluster center.

Similar to Fuzzy C-Means, MIFCM optimizes the objective function iteratively by updating the membership and cluster centers. The MIFCM membership value is updated using equation

$$\beta_{ij} = \mu_{ij} + \pi_{ij} \tag{21}$$

where μ_{ij} is the membership value and π_{ij} is the hesitation value. The membership value μ_{ij} is computed as follows:

$$\mu_{ij} = \frac{1}{\sum\limits_{k=1}^{c}\left(\frac{d_H(x_j,v_i)}{d_H(x_j,v_k)}\right)^{\frac{2}{m-1}}} \tag{22}$$

The hesitation value π_{ij} is the combination of the membership and the non-membership value, and it is computed as:

$$\pi_{ij} = 1 - \mu_{ij} - \eta_{ij} \tag{23}$$

where η_{ij} is the non-membership value. To compute the non- membership value, Sugeno's and Yager's intuitionistic fuzzy complement generators are used and the value is computed using Equations (24) and (25), respectively.

$$\eta_{ij} = \frac{1 - \mu_{ij}}{1 + \alpha\mu_{ij}} \tag{24}$$

$$\eta_{ij} = (1 - (\mu_{ij})^{\alpha})^{\frac{1}{\alpha}} \tag{25}$$

where $\alpha > 0$ is constant.

In this paper, we employed both Sugeno's and Yager's complement generators. MIFCM using Sugeno's function is named MIFCM_S and similarly MIFCM using Yager's function is named MIFCM_Y. Furthermore, cluster centers are updated using Equation (26).

$$v_i = \frac{\sum_{j=1}^{n} \beta_{ij}^m x_j}{\sum_{j=1}^{n} \beta_{ij}^m} \tag{26}$$

MIFCM is an iterative process, and it stops when the convergence criteria are satisfied (i.e., the difference between the objective function value of successive iterations is less than the user-specified stopping criteria value).

2.2.4. Consensus Clustering Using Voting Schema

In this section, the segmentation results are combined using voting schema. Let n be the number of voxels presented $X = \{x_1, x_2, x_3, \ldots\ldots, x_n\}$ and t be the set of clustering algorithms considered for clustering the n voxels, i.e., $\Pi = \{\pi_1, \pi_2.,,,, \pi_t\}$. Each clustering algorithm π_i maps x_i to c clusters. The problem of consensus clustering is to find a new π^* that best summarizes the clustering ensemble. In the proposed work, the input brain images voxels are segmented using the above-discussed four clustering algorithms. After convergence of each algorithm, each voxel is assigned to its corresponding cluster based on the membership value. Let U_1, U_2, U_3, and U_4 represent the membership matrix of RSKFCM, GSKFCM, MIFCM_S, and MIFCM_Y, respectively. From these membership matrices, a label for each pixel is computed. The pixel x_j is assigned a label of a cluster for which it has maximum membership value. Let P_1, P_2, P_3, and P_4 be the label matrix created for RSKFCM, GSKFCM, MIFCM_S, and MIFCM_Y, respectively. From these label matrices, consensus results are produced using a voting method. The pixel x_j is assigned to a cluster based on the maximum number of cluster labels, i.e., $label = argmax_i\left(P_{(l)}^{(i)}\right)$, where $l = \{1, 2, 3, 4\}$ and $1 \leq i \leq c$. Algorithm 1 presents the individual steps involved in the proposed method.

Algorithm 1: Consensus Clustering using voting scheme

Data: Input image $X = \{x_1, \ldots, x_j, \ldots, x_n\}$, Stopping criteria (ε), m, number of clusters C

Result: Segmentation results, Cluster centers

1 Obtain membership matrix for each clustering algorithm
2 Construct label matrix for each algorithm i.e P_1, P_2, P_3 and P_4
3 The pixel x_j is assigned to a cluster based on maximum number of cluster label, i.e., $label = argmax_i\left(P_{(l)}^{(i)}\right)$, where $l = \{1, 2, 3, 4\}$ and $1 \leq i \leq c$.
4 Update the cluster centers by considering new cluster assignments

3. Experimental Results

This section presents the dataset used for experimentation, the metrics used to evaluate the proposed method, and the experimental setup followed by results and discussion.

3.1. Datasets

To assess the proposed method, we carried out experiments on two publicly available standard datasets.

3.1.1. OASIS

The Open Access Series of Imaging Studies (OASIS), is a publicly available standard MRI dataset (See the "Open Access Series of Imaging Studies" (OASIS) project's web site at https://www.oasis-brains.org/ (accessed on 19 April 2022)). This dataset consists of 416 cross-sectional data from subjects aged between 18 and 96. The images in the dataset are of 1.25 mm thickness and of 256 × 256 × 128 resolution.

3.1.2. IBSR18

The Internet Brain Segmentation Repository (IBSR18) (See the "Internet Brain Segmentation Repository" (IBSR) project's web site at https://www.nitrc.org/projects/ibsr/ (accessed on 19 April 2022)) was created by the Center for Morphometic Analysis at the Massachusetts General Hospital. IBSR18 contains 18 T1 weighted MR brain images and their corresponding segmentation ground truth images. The images have 1.55 mm thickness with a resolution of 256 × 256 × 128. All the images are bias field corrected using the Autoseg method developed by the University of North Carolina at Chapel Hill (See the "AutoSeg" repository https://www.nitrc.org/projects/autoseg/ (accessed on 19 April 2022)).

3.2. Evaluation Metrics

Usually, the segmentation results are evaluated for CSF, GM, and WM tissues using the following three evaluation metrics: overlap measure, boundary measure, and volume measure. In this paper, we evaluate our proposed method using all three measures.

Dice similarity Coefficient (DC): Dice similarity coefficient [35] is used to estimate the spatial overlap between the ground truth and the segmentation results, using the following equation.

$$\mathrm{DC} = \frac{2 * |Seg\_Im \cap GT\_Im|}{|Seg\_Im| + |GT\_Im|} \tag{27}$$

where Seg_Im is the segmentation result of the proposed method, and GT_Im is the ground truth. Higher DC represents more accurate segmentation.

Hausdorff Distance (HD): The Hausdorff distance [36] is used as the boundary measure, and it is calculated between the ground truth points φ and the segmented points $\hat{\varphi}$ using the following equation:

$$\mathrm{HD} = \max_{\hat{\varphi} \in Seg\_Im} \min_{\varphi \in GT\_Im} |\hat{\varphi} - \varphi| \tag{28}$$

The original Hausdorff distance is affected by outliers [37]. Thus, to reduce the influence of outliers, we used the 95th percentile of the Hausdorff distance. In the following, therefore, HD refers to the 95th percentile of the Hausdorff distance, and lower HD represents a more accurate result.

Absolute Volume Difference (AVD): Absolute Volume Difference is a volume measure used to compute volume difference between the ground truth and the obtained results. It is computed as follows:

$$\mathrm{AVD} = \frac{|Seg\_Im| - |GT\_Im|}{|GT\_Im|} \tag{29}$$

Lower AVD indicates a more accurate segmentation.

3.3. Experimental Setup

In this paper, we set the fuzzifier m value as two, stopping criterion ε to 0.0001, and initialized cluster centers randomly. We used voxel intensity as a feature. We let the window size N_k vary in $\{3,5,7\}$. From the experiments, it is found that when $K = 3$, performance is better. Therefore, we set $K = 3$ in all the experiments. In addition, to set the value of α, we varied α from 0.1 to 1. From the experiments, it is found that when $\alpha = 0.9$ performance is better, and this value was used in all the experiments. To assess the performance of the proposed method, we used 10-fold cross validation. The proposed model was implemented and experimented in MatLab 2016a.

3.4. Results

This section presents the results on the OASIS and IBSR18 datasets. The performance of the proposed method is compared with state-of-the-art methods. In addition, the performance is also compared with the latest version of standard brain segmentation tools such as FSL[38], SPM12 [39] and FreeSurfer [40]. The following methods are considered for comparison:

- HMRF-EM [8]: This method combines hidden Markov random field (HMRF) with an EM algorithm for MRI image segmentation. The main advantage of this method is it derives how the spatial information is encoded through the mutual influences of neighboring sites.
- FAST-PVE [41]: This method uses Markov random field(MRF) for brain tissue segmentation. To increase the speed, this method uses fast iterated conditional modes to solve MRFs.
- MSSEG [42]: This method deal with images in the presence of WM lesions. This approach integrates a robust partial volume tissue segmentation with WM outlier rejection and filling, combining intensity and probabilistic and morphological prior maps.
- R-FCM [43]: This method models the intensity inhomogeneities as a gain field that causes image intensities to smoothly and slowly vary through the image space. It iteratively adapts to the intensity inhomogeneities and is completely automated.
- SFCM [44]: This method generalizes the objective function of a conventional FCM by incorporating a spatial penalty on the membership function.
- FANTASM [45]: This method is the extension of an adaptive FCM. In this method, an additional constraint is placed on the membership functions that force them to be spatially smooth.
- PVC [31]: This method uses a partial volume model for MRI brain tissue segmentation. First, it classifies nonbrain tissue using a combination of anisotropic diffusion filtering, edge detection, and mathematical morphology. Further, the local estimates are computed by fitting a partial volume tissue measurement model to histograms of neighborhoods about each estimate point. Voxels in the intensity-normalized image are then classified into six tissue types using a maximum a posteriori classifier.
- SPM5 [46]: This method is based on a mixture of Gaussians. In addition, it is extended to incorporate a smooth intensity variation and nonlinear registration with tissue probability maps.
- GAMIXTURE [47]: This method employs finite mixture models (FMMs) for brain tissue segmentation. To deal with FMM complex optimization, this method employs a global optimization algorithm that uses blended crossover and a new permutation operator.
- ANN [48]: This method is based on a self-organizing map (SOM). Initially, the feature vector is extracted from the intensity of the pixel and its n nearest neighbors. Further, to improve the robustness, statistical transformation is applied to the extracted feature vector. Finally, each pixel is classified using SOM.
- KNN [49]: This method uses K-NN for brain tissue segmentation.
- BrainSuit09 [50]: This is an automatic brain image analysis tool. The tool provides a sequence of low-level operations in a single package that can produce accurate brain segmentation in clinical time.
- SVPASEG [51]: This method uses local image models for brain tissue segmentation. This model combines the local models for tissue intensities and Markov Random Field (MRF) into a global probabilistic image model. Finally, the parameters for the local intensity models are obtained without supervision by maximizing the weighted likelihood of a certain finite mixture model.
- EGC-SOM [52]: This method uses self-organizing maps (SOM) for brain tissue segmentation. Initially, first- and second-order features are extracted using overlapping

windows. Further, evolutionary computing is used for feature selection. Finally, map units are grouped using SOM.

- RF-CRF [53]: This method uses a conditional random field with a random forest for brain tissue segmentation. This method uses intensities, gradients, probability maps, and locations as features.

3.4.1. Results on OASIS Dataset

The OASIS dataset contains the images which are already skull stripped. Bias field correction was performed using the ROBEX tool [34]. Figure 1 shows the qualitative segmentation results obtained using the proposed method. We compared the results of the proposed method with the three state-of-the-art methods, i.e., HMRF-EM [8], FAST-PVE [41], and MSSEG [42]. All three methods' codes are available on the authors' websites. The comparison of their results is presented in Table 1. We notice that the proposed model has better performance with regard to CSF, GM, and WM when compared to the other methods.

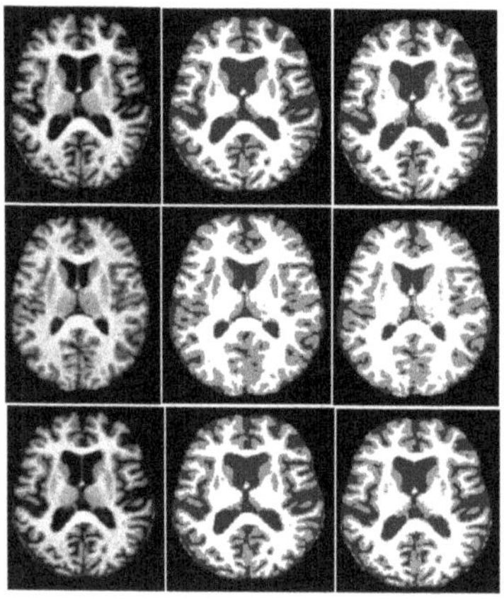

Figure 1. Segmentation results on OASIS dataset: first column, original image; second column, ground truth; and third column, segmentation result fused on ground truth.

Table 1. Results comparison with state-of-the-art methods on OASIS dataset (Mean ± std).

Method	CSF			GM			WM		
	DC	HD	AVD	DC	HD	AVD	DC	HD	AVD
HMRF-EM [8]	61.47 ± 2.32	7.17 ± 1.62	12.51 ± 8.57	79.65 ± 4.26	5.14 ± 1.62	4.11 ± 8.04	83.82 ± 4.02	5.09 ± 1.31	3.33 ± 8.64
FAST-PVE [41]	54.08 ± 3.61	7.17 ± 1.51	12.51 ± 7.28	78.97 ± 2.24	5.14 ± 0.92	4.11 ± 6.34	85.11 ± 2.61	5.09 ± 1.11	3.33 ± 6.24
FSL [38]	79.72 ± 3.64	4.78 ± 1.92	9.36 ± 5.32	87.84 ± 2.37	5.33 ± 0.81	5.37 ± 6.57	88.51 ± 2.31	5.13 ± 1.34	8.18 ± 4.95
SPM12 [39]	80.46 ± 4.02	6.71 ± 2.01	20.77 ± 6.04	89.52 ± 1.52	3.91 ± 0.76	3.67 ± 4.82	88.11 ± 2.42	4.54 ± 0.95	2.7 ± 5.24
FreeSurfer [40]	84.33 ± 3.96	4.4 ± 2.04	4.33 ± 4.06	91.47 ± 2.44	4.18 ± 0.59	2.63 ± 4.91	90.48 ± 1.31	3.8 ± 0.59	2.77 ± 5.64
MSSEG [42]	89.95 ± 1.62	4.18 ± 1.62	4.71 ± 1.62	91.24 ± 1.62	4.31 ± 1.62	2.87 ± 1.62	89.58 ± 1.62	4.39 ± 1.62	2.91 ± 1.62
RSKFCM [28]	90.06 ± 2.79	4.06 ± 1.61	4.31 ± 4.52	92.31 ± 2.61	4.21 ± 0.46	2.31 ± 3.42	90.51 ± 1.52	4.31 ± 0.42	2.81 ± 5.29
GSKFCM [29]	91.23 ± 2.82	4.08 ± 1.53	4.28 ± 3.26	92.51 ± 2.32	4.13 ± 0.42	2.11 ± 3.61	90.62 ± 1.61	4.28 ± 0.68	2.71 ± 5.41
MIFCM_S [20]	89.21 ± 3.02	4.23 ± 1.42	4.51 ± 3.14	89.81 ± 2.41	4.41 ± 0.31	2.97 ± 3.59	87.28 ± 1.71	4.59 ± 0.82	3.01 ± 5.16
MIFCM_Y [20]	92.65 ± 3.06	3.96 ± 1.46	3.91 ± 3.02	93.64 ± 2.51	4.23 ± 0.32	2.16 ± 3.42	92.61 ± 1.68	4.31 ± 0.84	2.17 ± 5.24
Proposed Method (consensus clustering)	**93.64** ± 2.15	**3.16** ± 1.31	**3.85** ± 2.06	**94.71** ± 2.30	**4.01** ± 0.21	**2.06** ± 2.96	**93.17** ± 1.32	**4.26** ± 0.81	**2.07** ± 4.21

3.4.2. Results on IBSR18 Dataset

The images in the IBSR18 are already bias field corrected. Hence, we have not applied any bias field correction technique. We conducted the experiments by removing the skull using a ground truth mask. Figure 2 shows the qualitative segmentation results obtained using the proposed method. The main limitation of the IBSR18 dataset is that it considers sulcal CSF as GM. The authors in [54] compared 10 existing methods without considering the sulcal CSF. Following [55,56], in our study we did not removed the sulcal CSF. We have compared the results of the proposed method with state-of-the-art methods. As all the considered methods have used DC alone as an evaluation metric, Table 2 shows the results only on the DC of the IBSR18 dataset. From this comparison, it is clear that the proposed model has better performance concerning CSF, GM, and WM when compared to the other methods.

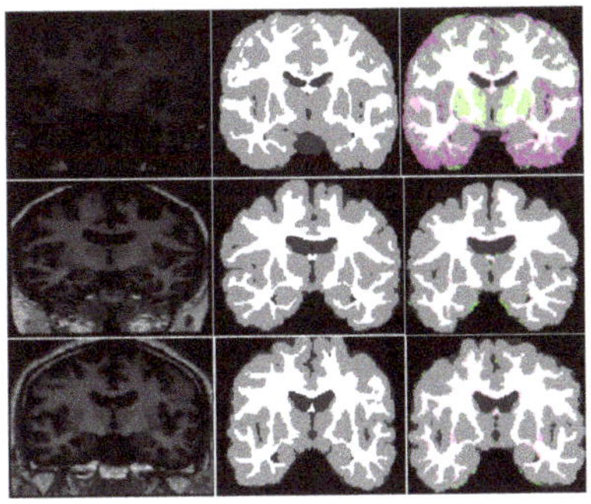

Figure 2. Segmentation results on IBSR18 dataset: first column, original image; second column, ground truth; and third column, segmentation result fused on ground truth.

Table 2. Result comparison with state-of-the-art methods on IBSR18 dataset only in terms of DC.

Method	GM		WM		CSF	
	mean	std	mean	std	mean	std
R-FCM [43]	65.00	0.05	75.00	0.05	NA	NA
NL-FCM [43]	72.00	0.05	74.00	0.05	NA	NA
FCM [43]	74.00	0.05	72.00	0.05	NA	NA
HMRF-EM [8]	74.60	0.04	89.60	0.02	12.60	0.05
SFCM [44]	70.60	0.06	86.60	0.03	16.60	0.07
FANTASM [45]	71.60	0.06	88.60	0.03	11.60	0.06
PVC [31]	70.60	0.08	83.60	0.07	13.60	0.06
SPM5 [46]	68.60	0.07	86.60	0.02	10.60	0.05
GAMIXTURE [47]	78.60	0.08	87.60	0.02	15.60	0.09
ANN [48]	70.60	0.07	87.60	0.03	11.60	0.06
KNN [49]	79.60	0.03	86.60	0.03	16.60	0.07
BrainSuit09 [50]	72.00	0.09	83.00	0.08	NA	NA
SVPASEG [51]	81.60	0.03	88.60	0.02	16.60	0.07
SPM8 [39]	81.60	0.02	88.60	0.01	17.60	0.08
EGC-SOM [52]	73.00	0.05	76.00	0.04	NA	NA
HFS-SOM [52]	60.00	0.09	60.00	0.08	NA	NA
FAST-PVE [41]	78.00	0.08	86.00	0.04	NA	NA
FAST-PVE(S-ICM) [41]	78.00	0.08	86.00	0.04	NA	NA
RF-CRF [53]	96.10	0.01	92.00	0.02	92.00	0.03
RF-CRF1 [53]	94.00	0.01	89.00	0.02	88.00	0.03
FSL [38]	78.13	0.04	85.94	0.13	75.02	0.04
FreeSurfer [40]	79.62	0.06	86.17	0.12	76.42	0.06
SPM12 [39]	82.30	0.04	89.82	0.02	78.62	0.14
RSKFCM [28]	96.68	0.09	93.55	0.10	93.41	0.08
GSKFCM [29]	96.72	0.03	93.58	0.02	93.43	0.02
MIFCM_S [20]	96.74	0.41	93.62	0.43	93.86	0.71
MIFCM_Y [20]	96.82	0.15	93.64	0.15	94.02	0.15
Proposed Method (consensus clustering)	**97.31**	**0.01**	**94.50**	**0.04**	**95.68**	**0.02**

3.5. Autism Spectrum Disorder Detection Using Proposed Method

Additionally, the proposed consensus clustering method has been evaluated on a practical autism spectrum disorder (ASD) detection problem. We used publicly available Autism Brain Imaging Data Exchange (ABIDE) data for this study. The ABIDE dataset contains 1112 subjects, 571 of them normal, and 531 of them with Autism Spectrum Disorders. We used 1054 of the 1112 subjects for this study, and the rest were rejected for improper segmentation using voxel-based morphometry (VBM). In this study, we employ a feature extraction method based on the VBM [57]. VBM is a fast and automatic method for determining the difference in gray matter structure between normal and and ASD patient brains [58]. In our VBM analysis, unified segmentation, smoothing, and statistical analysis were performed as preprocessing steps. In the unified segmentation step, tissue segmentation, bias correction, and image registration were combined in a single generative model [46]. The segmented and registered gray matter images were then smoothed by convolving with an isotropic Gaussian kernel. Here, a 10 mm full-width at half-maximum kernel was employed. A two-sample t-test was performed on the smoothed images, and gray matter volume was used as the covariate. This VBM analysis revealed significant gray matter volume increases in the normal persons in comparison with the ASD patients. The voxel location of significant regions were used as a mask. All segmented gray matter images were used to extract gray matter tissue probability values using a mask. A total of 989 features were obtained. and these were used as an input to the proposed method. Table 3 presents the performance comparison for Autism Spectrum Disorder Detection. The results of the proposed method are compared with traditional K-Means and variants of FCM methods. It is observed in Table 3 that the proposed method outperforms other methods.

Table 3. Performance comparison for Autism Spectrum Disorder Detection.

Method	Accuracy	Precision	Recall
K-Means	52.28 ± 2.35	0.531 ± 0.098	0.542 ± 0.077
FCM	52.36 ± 2.05	0.534 ± 0.081	0.546 ± 0.079
RSKFCM [28]	54.06 ± 1.21	0.548 ± 0.074	0.556 ± 0.068
GSKFCM [29]	54.61 ± 1.61	0.550 ± 0.061	0.557 ± 0.073
MIFCM_S [20]	55.08 ± 1.34	0.551 ± 0.067	0.559 ± 0.085
MIFCM_Y [20]	55.18 ± 1.27	0.556 ± 0.058	0.560 ± 0.054
Proposed Method (consensus clustering)	56.84 ± 1.09	0.565 ± 0.047	0.570 ± 0.049

3.6. Discussion

Brain images are very complex, largely uncertain, and imprecise. The fuzzy clustering based methods are capable of handling the aforementioned challenges.In this paper, we have combined the results from four variants of FCM clustering methods. The RSKFCM and GSKFCM are proven to be less sensitive to noise due to the use of kernel distance and the addition of neighborhood information. The MIFCM_S and MIFCM_Y are based on an intuitionistic fuzzy set which considers non-membership value along with membership value. Thus, in comparison to RSKFCM and GSKFCM, MIFCM methods handled the uncertainty better and achieved better results. Since we combined the advantages of all four clustering methods, our proposed consensus clustering method achieved better results compared to state-of-the-art methods.

On the OASIS dataset, the proposed method outperforms other methods in comparison. The OASIS dataset contains skull stripped T1 weighted MRI images. The main challenge in the OASIS dataset is the presence of WM lesions. The presence of WM lesions affects the overall segmentation accuracy of the proposed method. On the IBSR18 dataset, the proposed method outperforms all other methods in comparison. The images in the IBSR18 dataset are affected by acquisition artifacts which have direct impact on the WM tissue segmentation. On the other hand, lack of sulcal CSF labelling in the ground truth affects the GM and the CSF tissue segmentation results. Additionally, the proposed consensus clustering method has been evaluated on a practical autism spectrum disorder (ASD) detection problem. The proposed method outperforms other clustering algorithms. Even though the proposed consensus clustering algorithm is capable of handling noise and can exploit the spatial information in the image, it fails to capture the variations within the neighbourhood voxels. In addition, the time complexity of the proposed algorithm is more compared to individual clustering algorithms.

4. Conclusions

In this paper, a new approach for MRI Brain tissue segmentation is presented. The proposed method is based on the consensus clustering method. In consensus clustering, the results of four variants of fuzzy clustering methods are combined to achieve better results. The results of the proposed methods are evaluated using three performance metrics, i.e., DC, HD, and AVD. The competence of the proposed method is validated using two publicly available datasets: OASIS and IBSR18. From experimentation, it has turned out that our proposed method obtains the best result compared to other contemporary methods on the OASIS and IBSR18 datasets. Additionally, the proposed consensus clustering method has been evaluated on a practical autism spectrum disorder (ASD) detection problem.

Author Contributions: Methodology, S.V.A.K. and E.Y.; Validation, S.V.A.K.; Writing—original draft, S.V.A.K.; Writing—review & editing, E.Y. and H.P. All authors have read and agreed to the published version of the manuscript.

Funding: This research received no external funding.

Institutional Review Board Statement: Not applicable.

Informed Consent Statement: Not applicable.

Data Availability Statement: Not applicable.

Conflicts of Interest: The authors declare no conflict of interest.

References

1. Despotovic, I.; Goossens, B.; Philips, W. MRI segmentation of the human brain: Challenges, methods, and applications. *Comput. Math. Methods Med.* **2015**, *2015*, 450341. [CrossRef] [PubMed]
2. Zhang, Y.; Li, Y.; Kong, Y.; Wu, J.; Yang, J.; Shu, H.; Coatrieux, G. GSCFN: A graph self-construction and fusion network for semi-supervised brain tissue segmentation in MRI. *Neurocomputing* **2021**, *455*, 23–37. [CrossRef]
3. Veluchamy, M.; Subramani, B. Brain tissue segmentation for medical decision support systems. *J. Ambient Intell. Humaniz. Comput.* **2021**, *12*, 1851–1868. [CrossRef]
4. Song, J.; Yuan, L. Brain tissue segmentation via non-local fuzzy c-means clustering combined with Markov random field. *Math. Biosci. Eng.* **2022**, *19*, 1891–1908. [CrossRef]
5. Coleman, G.B.; Andrews, H.C. Image segmentation by clustering. *Proc. IEEE* **1979**, *67*, 773–785. [CrossRef]
6. Chen, C.W.; Luo, J.; Parker, K.J. Image segmentation via adaptive K-mean clustering and knowledge-based morphological operations with biomedical applications. *IEEE Trans. Image Process.* **1998**, *7*, 1673–1683. [CrossRef]
7. Wu, M.N.; Lin, C.C.; Chang, C.C. Brain tumor detection using color-based k-means clustering segmentation. In Proceedings of the IEEE Third International Conference on Intelligent Information Hiding and Multimedia Signal Processing (IIHMSP), Sendai, Japan, 26–28 November 2007; pp. 245–250.
8. Zhang, Y.; Brady, M.; Smith, S. Segmentation of brain MR images through a hidden Markov random field model and the expectation-maximization algorithm. *IEEE Trans. Med. Imaging* **2001**, *20*, 45–57. [CrossRef] [PubMed]
9. Phillips, W.; Velthuizen, R.; Phuphanich, S.; Hall, L.; Clarke, L.; Silbiger, M. Application of fuzzy c-means segmentation technique for tissue differentiation in MR images of a hemorrhagic glioblastoma multiforme. *Magn. Reson. Imaging* **1995**, *13*, 277–290. [CrossRef]
10. Kong, J.; Wang, J.; Lu, Y.; Zhang, J.; Li, Y.; Zhang, B. A novel approach for segmentation of MRI brain images. In Proceedings of the IEEE Mediterranean Electrotechnical Conference (MELECON), Lemesos, Cyprus, 18–20 April 2016; pp. 525–528.
11. Ahmed, M.N.; Yamany, S.M.; Mohamed, N.; Farag, A.A.; Moriarty, T. A modified fuzzy c-means algorithm for bias field estimation and segmentation of MRI data. *IEEE Trans. Med. Imaging* **2002**, *21*, 193–199. [CrossRef]
12. Liew, A.W.C.; Yan, H. An adaptive spatial fuzzy clustering algorithm for 3-D MR image segmentation. *IEEE Trans. Med. Imaging* **2003**, *22*, 1063–1075. [CrossRef]
13. Wang, J.; Kong, J.; Lu, Y.; Qi, M.; Zhang, B. A modified FCM algorithm for MRI brain image segmentation using both local and non-local spatial constraints. *Comput. Med. Imaging Graph.* **2008**, *32*, 685–698. [CrossRef] [PubMed]
14. Zhang, D.Q.; Chen, S.C. Clustering incomplete data using kernel-based fuzzy c-means algorithm. *Neural Process. Lett.* **2003**, *18*, 155–162. [CrossRef]
15. Lin, K.P. A novel evolutionary kernel intuitionistic fuzzy *c*-means clustering algorithm. *IEEE Trans. Fuzzy Syst.* **2014**, *22*, 1074–1087. [CrossRef]
16. Atanassov, K.T. Intuitionistic fuzzy sets. *Fuzzy Sets Syst.* **1986**, *20*, 87–96. [CrossRef]
17. Iakovidis, D.K.; Pelekis, N.; Kotsifakos, E.; Kopanakis, I. Intuitionistic fuzzy clustering with applications in computer vision. In Proceedings of the International Conference on Advanced Concepts for Intelligent Vision Systems, Juan-les-Pins, France, 20–24 October 2008; Springer: Berlin, Germany, 2008; pp. 764–774.
18. Chaira, T. A novel intuitionistic fuzzy C means clustering algorithm and its application to medical images. *Appl. Soft Comput.* **2011**, *11*, 1711–1717. [CrossRef]
19. Kumar, S.A.; Harish, B.; Aradhya, V.M. A picture fuzzy clustering approach for brain tumor segmentation. In Proceedings of the 2016 Second International Conference on Cognitive Computing and Information Processing (CCIP), Mysuru, India, 12–13 August 2016; pp. 1–6.
20. Kumar, S.A.; Harish, B. A Modified intuitionistic fuzzy clustering algorithm for medical image segmentation. *J. Intell. Syst.* **2018**, *27*, 593–607. [CrossRef]

21. Verma, H.; Agrawal, R.; Sharan, A. An improved intuitionistic fuzzy c-means clustering algorithm incorporating local information for brain image segmentation. *Appl. Soft Comput.* **2016**, *46*, 543–557. [CrossRef]
22. Pedrycz, W.; Rai, P. Collaborative clustering with the use of Fuzzy C-Means and its quantification. *Fuzzy Sets Syst.* **2008**, *159*, 2399–2427. [CrossRef]
23. Punera, K.; Ghosh, J. Consensus-based ensembles of soft clusterings. *Appl. Artif. Intell.* **2008**, *22*, 780–810. [CrossRef]
24. Sevillano, X.; Alías, F.; Socoró, J.C. BordaConsensus: A new consensus function for soft cluster ensembles. In Proceedings of the 30th Annual International ACM SIGIR Conference on Research and Development in Information Retrieval, Amsterdam, The Netherlands, 23–27 July 2007; pp. 743–744.
25. Crespo, F.; Weber, R. A methodology for dynamic data mining based on fuzzy clustering. *Fuzzy Sets Syst.* **2005**, *150*, 267–284. [CrossRef]
26. Pedrycz, W. A dynamic data granulation through adjustable fuzzy clustering. *Pattern Recognit. Lett.* **2008**, *29*, 2059–2066. [CrossRef]
27. Wu, J.; Wu, Z.; Cao, J.; Liu, H.; Chen, G.; Zhang, Y. Fuzzy Consensus Clustering With Applications on Big Data. *IEEE Trans. Fuzzy Syst.* **2017**, *25*, 1430–1445. [CrossRef]
28. Kumar, S.A.; Harish, B. Segmenting mri brain images using novel robust spatial kernel fcm (rskfcm). In Proceedings of the Eighth International Conference on Image and Signal Processing, Cherbourg, France, 30 June–2 July 2014; pp. 38–44.
29. Kumar, S.A.; Harish, B.; Shivakumara, P. A novel fuzzy clustering based system for medical image segmentation. *Int. J. Comput. Intell. Stud.* **2018**, *7*, 33–66. [CrossRef]
30. Smith, S.M. Fast robust automated brain extraction. *Hum. Brain Mapp.* **2002**, *17*, 143–155. [CrossRef] [PubMed]
31. Shattuck, D.W.; Sandor-Leahy, S.R.; Schaper, K.A.; Rottenberg, D.A.; Leahy, R.M. Magnetic resonance image tissue classification using a partial volume model. *NeuroImage* **2001**, *13*, 856–876. [CrossRef] [PubMed]
32. Mikheev, A.; Nevsky, G.; Govindan, S.; Grossman, R.; Rusinek, H. Fully automatic segmentation of the brain from T1-weighted MRI using Bridge Burner algorithm. *J. Magn. Reson. Imaging* **2008**, *27*, 1235–1241. [CrossRef]
33. Sadananthan, S.A.; Zheng, W.; Chee, M.W.; Zagorodnov, V. Skull stripping using graph cuts. *NeuroImage* **2010**, *49*, 225–239. [CrossRef]
34. Iglesias, J.E.; Liu, C.Y.; Thompson, P.M.; Tu, Z. Robust brain extraction across datasets and comparison with publicly available methods. *IEEE Trans. Med. Imaging* **2011**, *30*, 1617–1634. [CrossRef]
35. Dice, L.R. Measures of the amount of ecologic association between species. *Ecology* **1945**, *26*, 297–302. [CrossRef]
36. Huttenlocher, D.P.; Klanderman, G.A.; Rucklidge, W.J. Comparing images using the Hausdorff distance. *IEEE Trans. Pattern Anal. Mach. Intell.* **1993**, *15*, 850–863. [CrossRef]
37. Mendrik, A.M.; Vincken, K.L.; Kuijf, H.J.; Breeuwer, M.; Bouvy, W.H.; De Bresser, J.; Alansary, A.; De Bruijne, M.; Carass, A.; El-Baz, A.; et al. MRBrainS challenge: Online evaluation framework for brain image segmentation in 3T MRI scans. *Comput. Intell. Neurosci.* **2015**, *2015*, 1. [CrossRef] [PubMed]
38. Woolrich, M.W.; Jbabdi, S.; Patenaude, B.; Chappell, M.; Makni, S.; Behrens, T.; Beckmann, C.; Jenkinson, M.; Smith, S.M. Bayesian analysis of neuroimaging data in FSL. *Neuroimage* **2009**, *45*, S173–S186. [CrossRef] [PubMed]
39. Ashburner, J.; Barnes, G.; Chen, C.; Daunizeau, J.; Flandin, G.; Friston, K.; Kiebel, S.; Kilner, J.; Litvak, V.; Moran, R. *SPM8 Manual*; Wellcome Trust Centre for Neuroimaging Institute of Neurology; UCL: London, UK, 2012.
40. Dale, A.M.; Fischl, B.; Sereno, M.I. Cortical surface-based analysis: I. Segmentation and surface reconstruction. *Neuroimage* **1999**, *9*, 179–194. [CrossRef] [PubMed]
41. Tohka, J. FAST-PVE: Extremely fast Markov random field based brain MRI tissue classification. In *Proceedings of the Scandinavian Conference on Image Analysis*; Springer: Berlin, Germany, 2013; pp. 266–276.
42. Valverde, S.; Oliver, A.; Roura, E.; González-Villà, S.; Pareto, D.; Vilanova, J.C.; Ramio-Torrenta, L.; Rovira, A.; Llado, X. Automated tissue segmentation of MR brain images in the presence of white matter lesions. *Med. Image Anal.* **2017**, *35*, 446–457. [CrossRef]
43. Pham, D.L.; Prince, J.L. Adaptive fuzzy segmentation of magnetic resonance images. *IEEE Trans. Med. Imaging* **1999**, *18*, 737–752. [CrossRef]
44. Pham, D.L. Spatial models for fuzzy clustering. *Comput. Vis. Image Underst.* **2001**, *84*, 285–297. [CrossRef]
45. Pham, D.L. Robust fuzzy segmentation of magnetic resonance images. In Proceedings of the 14th IEEE Symposium on Computer-Based Medical Systems (CBMS), Bethesda, ML, USA, 26–27 July 2001; pp. 127–131.
46. Ashburner, J.; Friston, K.J. Unified segmentation. *Neuroimage* **2005**, *26*, 839–851. [CrossRef]
47. Tohka, J.; Krestyannikov, E.; Dinov, I.D.; Graham, A.M.; Shattuck, D.W.; Ruotsalainen, U.; Toga, A.W. Genetic algorithms for finite mixture model based voxel classification in neuroimaging. *IEEE Trans. Med. Imaging* **2007**, *26*, 696–711. [CrossRef]
48. Tian, D.; Fan, L. A brain MR images segmentation method based on SOM neural network. In Proceedings of the the 1st IEEE International Conference on Bioinformatics and Biomedical Engineering (ICBBE), Wuhan, China, 6–8 July 2007; pp. 686–689.
49. De Boer, R.; Vrooman, H.A.; Van Der Lijn, F.; Vernooij, M.W.; Ikram, M.A.; Van Der Lugt, A.; Breteler, M.M.; Niessen, W.J. White matter lesion extension to automatic brain tissue segmentation on MRI. *Neuroimage* **2009**, *45*, 1151–1161. [CrossRef]
50. Shattuc, D.; Leahy, R. BrainSuite: An automated cortical surface identification tool. *Med. Image Anal.* **2002**, *6*, 129–142. [CrossRef]
51. Tohka, J.; Dinov, I.D.; Shattuck, D.W.; Toga, A.W. Brain MRI tissue classification based on local Markov random fields. *Magn. Reson. Imaging* **2010**, *28*, 557–573. [CrossRef] [PubMed]

52. Ortiz, A.; Gorriz, J.; Ramirez, J.; Salas-Gonzalez, D.; Llamas-Elvira, J.M. Two fully-unsupervised methods for MR brain image segmentation using SOM-based strategies. *Appl. Soft Comput.* **2013**, *13*, 2668–2682. [CrossRef]
53. Pereira, S.; Pinto, A.; Oliveira, J.; Mendrik, A.M.; Correia, J.H.; Silva, C.A. Automatic brain tissue segmentation in MR images using Random Forests and Conditional Random Fields. *J. Neurosci. Methods* **2016**, *270*, 111–123. [CrossRef] [PubMed]
54. Valverde, S.; Oliver, A.; Cabezas, M.; Roura, E.; Llado, X. Comparison of 10 brain tissue segmentation methods using revisited IBSR annotations. *J. Magn. Reson. Imaging* **2015**, *41*, 93–101. [CrossRef] [PubMed]
55. Yi, Z.; Criminisi, A.; Shotton, J.; Blake, A. Discriminative, semantic segmentation of brain tissue in MR images. *Med. Image Comput. Comput. Assist. Interv.* **2009**, *12*, 558–565. [PubMed]
56. Yaqub, M.; Javaid, M.K.; Cooper, C.; Noble, J.A. Investigation of the role of feature selection and weighted voting in random forests for 3-D volumetric segmentation. *IEEE Trans. Med. Imaging* **2014**, *33*, 258–271. [CrossRef]
57. Vigneshwaran, S.; Suresh, S.; Sundararajan, N.; Mahanand, B.S. Accurate detection of autism spectrum disorder from structural MRI using extended metacognitive radial basis function network. *Expert Syst. Appl.* **2015**, *42*, 8775–8790.
58. Ashburner, J.; Friston, K.J. Voxel-Based Morphometry-The Methods. *NeuroImage* **2000**, *11*, 805–821. [CrossRef]

applied sciences

MDPI

Article

Comparison of Bone Segmentation Software over Different Anatomical Parts

Claudio Belvedere [1,*], Maurizio Ortolani [1], Emanuela Marcelli [2], Barbara Bortolani [2], Katsiaryna Matsiushevich [1], Stefano Durante [3], Laura Cercenelli [2] and Alberto Leardini [1]

1 Movement Analysis Laboratory, IRCCS Istituto Ortopedico Rizzoli, 40136 Bologna, Italy; maurizio.ortolani@ior.it (M.O.); k.matsiushevich@gmail.com (K.M.); leardini@ior.it (A.L.)
2 eDIMES Lab—Laboratory of Bioengineering, Department of Experimental, Diagnostic and Specialty Medicine (DIMES), University of Bologna, 40138 Bologna, Italy; emanuela.marcelli@unibo.it (E.M.); barbara.bortolani@unibo.it (B.B.); laura.cercenelli@unibo.it (L.C.)
3 Management of Health Professions, IRCCS S. Orsola-Malpighi Hospital, 40138 Bologna, Italy; stefano.durante@aosp.bo.it
* Correspondence: belvedere@ior.it; Tel.: +39-051-636-6570; Fax: +39-051-636-6561

Abstract: Three-dimensional bone shape reconstruction is a fundamental step for any subject-specific musculo-skeletal model. Typically, medical images are processed to reconstruct bone surfaces via slice-by-slice contour identification. Freeware software packages are available, but commercial ones must be used for the necessary certification in clinics. The commercial software packages also imply expensive hardware and demanding training, but offer valuable tools. The aim of the present work is to report the performance of five commercial software packages (Mimics®, Amira™, D2P™, Simpleware™, and Segment 3D Print™), particularly the time to import and to create the model, the number of triangles of the mesh, and the STL file size. DICOM files of three different computed tomography scans from five different human anatomical areas were utilized for bone shape reconstruction by using each of these packages. The same operator and the same hosting hardware were used for these analyses. The computational time was found to be different between the packages analyzed, probably because of the pre-processing implied in this operation. The longer "time-to-import" observed in one software is likely due to the volume rendering during uploading. A similar number of triangles per megabyte (approximately 20 thousand) was observed for the five commercial packages. The present work showed the good performance of these software packages, with the main features being better than those analyzed previously in freeware packages.

Keywords: DICOM; image segmentation; bone models; STL file; musculo-skeletal modeling; additive manufacturing; 3D modeling

Citation: Belvedere, C.; Ortolani, M.; Marcelli, E.; Bortolani, B.; Matsiushevich, K.; Durante, S.; Cercenelli, L.; Leardini, A. Comparison of Bone Segmentation Software over Different Anatomical Parts. *Appl. Sci.* **2022**, *12*, 6097. https://doi.org/10.3390/app12126097

Academic Editors: Ioannis A. Kakadiaris, Michalis Vrigkas and Christophoros Nikou

Received: 22 April 2022
Accepted: 10 June 2022
Published: 15 June 2022

1. Introduction

Three-dimensional (3D) reconstruction of bone models is fundamental for musculo-skeletal biomechanics, particularly for subject-specific modeling [1–4]. Exact 3D bone morphology is becoming essential in orthopedics for the custom design and surgical planning of joint replacements [5–8]. In this context, also the recent large progress in 3D printing is contributing to the huge number of exploitations in orthopedics and traumatology, since this additive technology enables the cheap manufacturing of custom-made prostheses and implants, along with relevant cutting jigs, designed over the exact dimension, shape, and alignment of bone and joint defects starting from patient-specific anatomy [9–15].

The full process from medical images to final implants also allows physical replica of patient anatomy, valuable for pre-operative planning, surgical team training, and physician-to-patient communication, in addition, of course, to musculo-skeletal and finite element modeling [16–18]. Typically, medical images from computed tomography (CT) are processed to be segmented, i.e., to reconstruct bone surface mesh via slice-by-slice bone contour identification [19–22]. Image segmentation is a long and critical process, which implies manual, automatic, or semi-automatic tracking of the silhouette contours of the bony structures, and, therefore, requires anatomical knowledge, computer skills, and awareness of the scopes [15,23,24]. Many dedicated software packages are offered on the market, from freeware tools with basic functions [25–27], running more likely on fairly performing computers, to expensive software packages with more effective segmentation algorithms and features [26,28,29], likely running on powerful computers. The optimal image segmentation software should support the user in carefully defining the bone models, and eventually providing a file to be exported in standard stereo-lithography (STL) format, with a suitable number of triangles, a uniform mesh, and a minimum overall size. The reconstruction of 3D bone models requires extensive work; a good compromise should be found for each application, between the automation of the segmentation process and the quality of the final results [25,30–32].

The current commercial software packages for image segmentation claim high performance and valuable technical tools, but require robust hardware, demanding training, and careful maintenance, and thus these result in expensive licenses. Hence, cheap and easy-to-use software tools [33] are still pursued and utilized. The performance of a number of freeware software programs was previously analyzed and compared while processing fifteen different human bones from five different anatomical areas; a number of valuable features and fair quality of the reconstructed bone models were found [25]. However, large differences in the number of triangles of the output meshes and in the file size were found, with the triangles per megabyte (MByte) ratio ranging from around 4 to 20 thousand [25]. Distance map analysis amongst outputs from these different free software packages revealed that root-mean-square deviations ranged from 0.13 to 2.21 mm when averaged over the five anatomical areas [25].

However, the major concern of these freeware software packages is the lack of certification as medical diagnostic devices, i.e., the official recognition to be used as appropriate preoperative software for implant design and surgical planning in the standard clinical practice [34]. Hence, the aim of the present work is to report on the performance of five commercial image segmentation packages. For a possible reasonable comparison, also the same exact CT scans of bony parts that we previously examined with freeware software packages [25] were used. From the presented original combination of these two analyses, advantages and disadvantages of commercial and freeware software packages for bone segmentation from CT scans can be established.

2. Methods

2.1. CT Scan Collection

Medical images in Digital Imaging and Communications in Medicine (DICOM) format from three different subjects were taken from a previous work [25]. In detail, for each of them, a number of anatomical complexes of the upper and the lower limb, i.e., the shoulder, elbow, and wrist for the former, and the knee and ankle for the latter, were analyzed. These fifteen scans were from CT ('Brilliance 16-slice scanner', Philips Medical Systems; Best, The Netherlands), with matrix size 512 × 512, voxel size 0.29 × 0.29 × 0.8 mm, layer thickness 0.5 mm, and field of view and data collection protocol set according to the specific anatomical complex to be analyzed. These technical parameters and the size of the anatomical complexes under analysis resulted in the following numbers of images, for each of the three subjects: 365, 256, and 321 for the shoulder; 353, 299, and 282 for the elbow; 239, 212, and 319 for the wrist; 204, 299, and 241 for the knee; 290, 481, and 315 for the ankle.

2.2. Segmentation Software Packages

Each dataset was analyzed with five commercial software packages for medial image segmentation (Table 1), i.e., all those available in the local area where the present analysis was performed: (1) Mimics™ Innovation Suite (v. 24.0, Materialise Inc., Leuven, Belgium), (2) Amira™ (v. 2019.4, Thermo Fisher Scientific Inc., Waltham, MA,

USA), (3) D2P™ (DICOM-to-PRINT, v. 1.0.2.2055, 3D Systems Inc., Rock Hill, SC, USA), (4) Simpleware™ (v. 2021.06, Synopsys Inc., Mountain View, CA, USA), and (5) Segment 3D Print™ (v. 3.3 R 9056, Medviso AB, Lund, Sweden). Relevant technical requirements for each software program are reported in Table 1.

Table 1. Technical requirements for each software package analyzed (* for best performance, multiple hard drive configuration—3 or more HDDs or SSDs—in RAID 5 mode is recommended, as reported in relevant user manual).

	Mimics (v. 24.0)	**Amira (v. 2019.4)**	**D2P (v. 1.0.2.2055)**	**Simpleware (v. 2021.06)**	**Segment 3D Print (v. 3.3 R 9056)**
Recommended Processor	Intel Core i7 or equivalent	Intel64/AMD64 architecture	Intel Core i7	Intel Core i7 or equivalent	Any processor supporting CUDA-enabled graphics
Minimum RAM [GB]	4	2	16	16	16
Minimum HDD space [GB]	5	Not reported *	500	100	5
Supported Operating System	Windows 10 Pro/Enterprise version 1803, 1809, 1903, 1909, 2009 (64-bit) or Windows Server 2019 Standard version 10.0,	Windows 7/8/10 (64-bit) Linux x86_64 (64-bit): CentOS 7 Mac OS X High Sierra (10.13) and Mac OS X Mojave (10.14)	Windows 7 or 10 (64 bit)	Windows 10/Windows Server 2016 Linux *: - RHEL 7.x and 8.x - CentOS 7.x and 8.x	Windows 10 (64 bit)

HDD = Hard Disk Drive; **SSD** = Solid-State Drive; **RDP** = Remote Desktop Protocol; **CUDA** = Compute Unified Device Architecture; **RAID** = Redundant Array of Independent Disks.

2.3. Medical Image Segmentation Process

All analyses were executed on the same computer (platform Intel® Xeon W-2123 CPU @ 3.60GHz, 64 bit, 32 GB RAM; graphics card: NVIDIA GeForce RTX 2070 SUPER) and operating system (Windows 10, Microsoft Co., Redmond, WA, USA).

The overall operational segmentation principle was very similar over each software package (Figure 1): importing the DICOM files in the three anatomical projections, followed by image segmentation using the most suitable tools offered by the software, including thresholding and minor possible manual corrections to remove isolated voxel areas. The masks defined by image segmentation embedded all bones of the overall joints. Although editing after image segmentation was allowed in all five software packages, no additional shaping or meshing tools were used. By means of built-in functions to convert the segmented masks into surface meshes, the 3D models of the bones were generated and exported in STL format files, all in binary code and in little-endian mode.

A radiographer, with 4 years of experience in the radiological department of an orthopedic center and 4 years of experience in 3D bone model segmentation, performed all reconstructions, using the same computer to remove possible bias; this was the same radiographer who performed a similar analysis in a previous study [25]. The necessary technical support for using the selected five software packages was provided by relevant computer scientists. All the present 3D bone reconstructions with the five packages were concentrated over a period of time of one month.

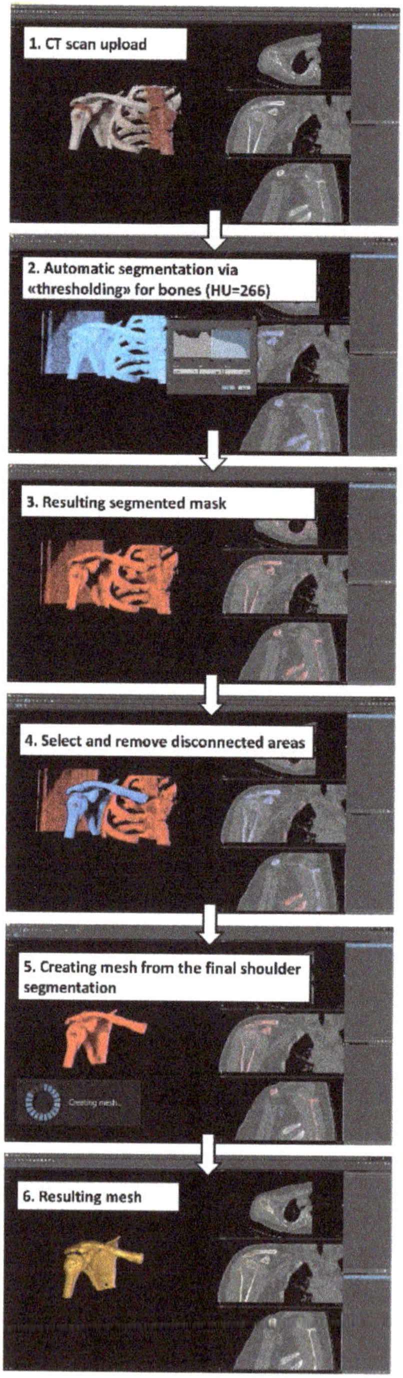

Figure 1. A diagram for an automatic/semi-automatic workflow for the segmentation process: present exemplary screenshots obtained during this process using D2P software; a very similar workflow was followed for the other four software packages.

2.4. Data Collection and Processing

During the image segmentation phase, the following parameters were collected for each of the software programs utilized: DICOM time to import, time to create the model, number of model triangles, model file size, and number of model triangles per megabyte. These are reported in terms of mean ± standard deviation over the five anatomical complexes and the three subjects, along with range values (min–max).

Furthermore, the Pearson product–moment correlation coefficient (R) was also used to derive correlations between the mean number of triangles and the mean file size for the models obtained using the commercial software packages analyzed in the present study and also for those previously obtained with freeware software packages [25]. Corresponding p-values are reported for assessing significance, this being accepted at $p < 0.05$.

3. Results

The software features considered in the present analysis are those reported in Table 2.

Table 2. Main features of the five software packages analyzed. The results are means ± standard deviation over the fifteen models analyzed (five anatomical areas, each from three subjects), along with range values (min–max). For the sake of comparison, the table format is similar to that reported in the previous work [25] (with the exception of information related to basic two-dimensional and 3D features, which is not reported here).

	Mimics (v.24.0)	Amira (v. 2019.4)	D2P (v. 1.0.2.2055)	Simpleware (v. 2021.06)	Segment 3D Print (v3.3 R 9056)
Time to import [s]	1.4 ± 0.5 *(1–2)*	2.4 ± 1.5 *(1–5)*	2.1 ± 0.3 *(2–3)*	2.5 ± 0.7 *(1–4)*	3.7 ± 1.1 *(2–6)*
Time to create the model [s]	5.8 ± 3.9 *(2–14)*	2.1 ± 0.4 *(2–3)*	11.1 ± 4.3 *(4–19)*	5.2 ± 2.4 *(3–10)*	23.9 ± 13.3 *(9–55)*
Number of triangles	849,995 ± 633,670 *(203,616–2,219,446)*	1,782,831.6 ± 1,145,476 *(532,574–3,843,000)*	1,752,240 ± 1,120,912 *(526,460–3,764,380)*	1,796,269 ± 1,132,502 *(568,436–3,834,908)*	1,816,860 ± 1,107,694 *(576,532–4,200,338)*
File size [megabytes]	76.0 ± 48.8 *(23.9–163)*	84.8 ± 54.5 *(25.3–183)*	83.4 ± 53.3 *(25.1–179)*	85.5 ± 53.9 *(27.1–182)*	86.5 ± 52.7 *(27.4–200)*
Number of triangles per MByte	10,433.4 ± 2111.5 *(6650–13616)*	21,019.2 ± 51.2 *(20,973–21,165)*	21,004.8 ± 32.0 *(20,971–21,077)*	21,003.1 ± 48.2 *(20,889–21,088)*	21,005.5 ± 31.1 *(20,972–21,079)*

Additional features were analyzed, but because these were found to be available exactly in each package, these are not reported in Table 2 but rather listed here below: unlimited number of image slices; multiplanar visualization and representation; correspondence amongst the coronal, axial, and sagittal orientations; crop and zoom; contrast and brightness adjustments; separation of the regions of interest; linear, angular, and volumetric measures; simultaneous planar images and 3D rendering; export of images and mesh data; export in STL file format.

The computational import time of the DICOM files was found to be within the range of 1–6 s considering all analyzed software packages. The longer "time-to-import" found for the Segment 3D Print software (3.7 ± 1.1 s) is likely due to the simultaneous creation of 3D volume rendering during the uploading process. Marked differences were observed in the time to create the model, this ranging from 2.1 ± 0.4 s in Amira to 23.9 ± 13.3 s in Segment 3D Print, on average.

Operability was satisfactory for each of the five software packages; these were efficient enough to obtain final results for all 15 anatomical models in a few hours (Figure 2). By keeping their own default settings in each of the five software packages, Mimics showed the smallest final number of triangles on average over the 15 models, i.e., around half that of the other software packages, which resulted eventually in the smallest and most consistent file size in terms of standard deviation (Table 2). At the end, a very similar ratio of triangles per MByte was observed, approximately 20 thousand, apart from Mimics, where this value was halved. In addition, the corresponding standard deviations reported in % of the mean values reveal a value of approximately 20% for Mimics, and less than 1% in the other software packages.

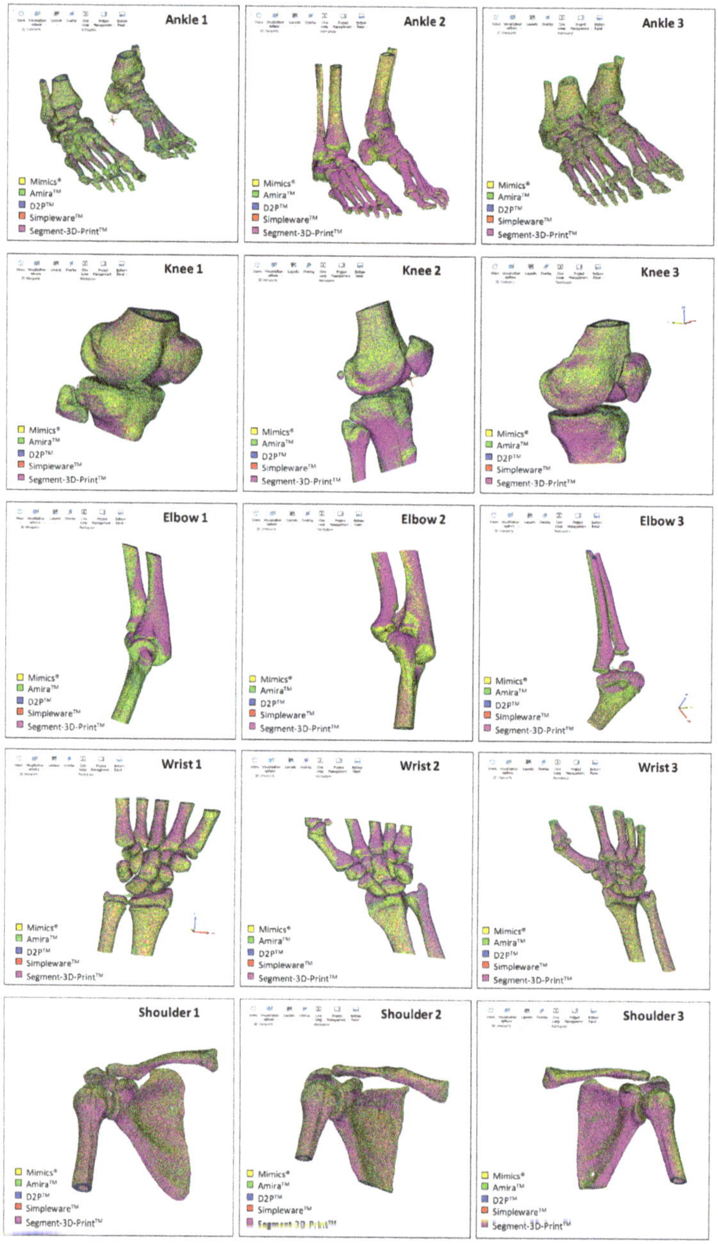

Figure 2. Superimposition of the resulting meshes from the five commercial software packages investigated; this is shown for each of the 15 DICOM files (three subjects for each of the five different anatomical areas). The STL models shown here for rough comparison were imported into Mimics for the present representation.

4. Discussion

For the first time, the main features of five commercial software packages (Mimics®, Amira™, D2P™, Simpleware™, and Segment 3D Print™) for the generation of STL bone models were investigated. Relevant performance in bone model reconstruction was analyzed by the same operator, using a single computer workstation, and compared accurately in fifteen different CT scans, from five anatomical areas with distinct morphological complexity, by the same operator and using the same computer hardware to avoid possible bias. Amongst the scope, there was also the quantitative comparison of specific features with those from four freeware software packages (3DSlicer, ITK-SNAP, InVesalius, and VuePACS3D—the latter being accessible free-of-charge in our radiological unit), analyzed recently by the same authors [25]. To our knowledge, these nine software packages are amongst the most popular for medical imaging segmentation of the musculo-skeletal system [22,27,28,35]. To make rational and objective comparisons, both the CT scans and the operator were the same in the two studies, and manual intervention was limited as much as possible also in the present analysis. Additionally, all software packages were used on the same computer to avoid the situation wherein segmentation performance is affected by the hardware specifications.

The present work did not seek to investigate either the segmentation algorithms, tools, and features of these commercial software packages, clearly very different from one another, or the degree of automation in 3D model reconstruction, but rather to assess only the main quantitative features and the gross results for comparison. Therefore, apart from a few basic final refinements, only the standard segmentation tools, such as thresholding, were used in each software package for bone shape reconstruction.

The present work obviously has limitations. To limit the cost of the present exercise, access to these commercial software packages was sought in the geographical area of the authors; five amongst the most popular were found and tested with no additional charges. Clearly, many other software packages are available on the market, but it was not possible to further enlarge this exercise, also because the condition of a single operator and a single computer was pursued. In this respect, the operator could not become very familiar with each software package in a short time; therefore, local users, already familiar with these, provided some support, only that necessary to obtain the final results. Threshold values for optimal mask visualization in bone segmentation were set by the operator in each software program, according to the specific density of each bone under analysis but generally in the range of 130–226 Hounsfield unit values, as in the previous work [25]. Of course, these commercial software packages feature many additional tools for manual editing and refinements, particularly with regard to the final number and the density of mesh triangles, but these were not exploited, to maintain the comparison of the initial basic performance. Finally, given the scope of this work, there was no need to distinguish between the different bones within a model, so all bones in each scan were segmented as a single object.

Our findings compare well with recent similar studies in the literature in terms of tested scan resolution, threshold, and accuracy of 3D bone model reconstruction [23,36]. With respect to similar previous studies where features and reconstruction techniques from different software packages are investigated and compared [22,27,28,35], the present work offers quantitative objective outcomes also in terms of the number of triangles, file size, and relevant ratio. In addition, the present analysis was not biased by single anatomical areas or limited scans, but involved full morphological reconstructions from five different anatomical areas with different complexity, overall from fifteen different subjects.

With respect to the freeware software analysis by the present authors [25], the segmentation thresholds and reconstruction algorithms were of course very different. In terms of the mean number of triangles, the four freeware programs were within the range of the commercial software packages, with Mimics® as the minimum (around 850,000) and Segment 3D Print™ as the maximum (around 1,750,000). The mean file size was found compatible with these differences, as expressed well in the triangles-per-MByte ratio (Table 2), which, apart from Mimics™, was approximately 20 thousand in the present commercial software packages and in the free 3D Slicer™ (Brigham and Women's Hospital, Boston, MA, USA) and InVesalius™ (Renato Archer Information Technology Center, Campinas, Brazil) of the previous paper; in ITK-SNAP™ (PICSL University of Pennsylvania, Philadelphia, PA, USA) and VuePACS3D (Carestream™, Rochester, NY, USA), this ratio was, respectively, 5.7 and 3.9 thousand, relatively closer to Mimics™. However, the ability to export the models both in American Standard Code for Information Interchange (ASCII) and binary STL files gives the option of more readable data for debugging and coding, or less space to store the same amount of data, respectively. Furthermore, it is very interesting to note (Figure 3) that for both the commercial and freeware software packages providing binary code for STL export, there is an overall linear trend between the file size and the number of triangles of the output mesh.

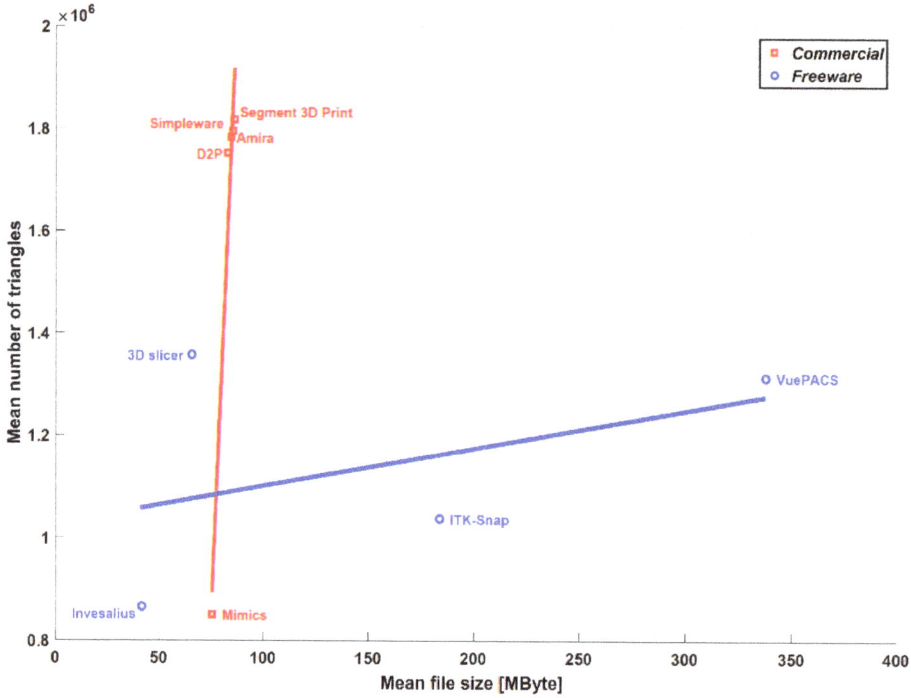

Figure 3. Graphical representation in terms of mean number of triangles and mean file size of the output meshes over the 15 anatomical models. Results obtained in the present study with commercial software are superimposed to the corresponding results previously obtained by these authors with freeware software on the same models [25]. Corresponding linear regression lines are superimposed for comparison.

Although this behavior is detectable in both the software types, only for the commercial ones is this statistically significant, the correlation coefficient being 0.98, with an associated p-value equal to 0.004. However, despite the features and the good performance of the freeware software packages for 3D bone model reconstruction reported recently by these authors [25], these software packages cannot currently be used worldwide in clinical practice because of the required certification as medical devices, according to the national regulations [34].

5. Conclusions

The present analysis has assessed five commercial software packages and found that the main features are better than those of freeware software packages, as expected.

Clearly, only the basic features of these commercial packages were evaluated in the present analysis, whereas their ancillary utility would be a matter for future studies. However, these features shall be assessed also together with other aspects, such as license conditions, costs, accessibility, ease-of-use, etc. Moreover, the intended use of the final 3D bone models should be considered, e.g., whether they are for finite element or musculo-skeletal modeling, prosthesis or custom jig design, clinical research, or medical education. Future relevant work shall compare the results obtained from traditional manual or semi-automatic segmentation tools with modern automatic segmentation software packages, as performed recently for other software packages [37,38].

Author Contributions: Conceptualization: C.B., E.M. and A.L.; methodology: C.B., S.D. and L.C.; software: M.O., K.M. and B.B.; validation: M.O., K.M. and B.B.; formal analysis: C.B., S.D. and L.C.; investigation: C.B., E.M. and A.L.; resources: A.L., S.D. and E.M.; writing—original draft preparation: C.B. and A.L.; writing—review and editing: C.B., M.O., E.M., B.B., K.M., S.D., L.C. and A.L.; visualization: M.O., K.M., B.B. and L.C.; supervision: C.B. and A.L.; project administration: A.L.; funding acquisition: A.L. All authors have read and agreed to the published version of the manuscript.

Funding: This study was funded by the Italian Ministry of Economy and Finance, program "5 per mille".

Informed Consent Statement: Patient consent was waived since DICOM data came from a local DICOM repository and were provided in fully anonymized form.

Data Availability Statement: The authors confirm that the data supporting the results of this study are available within the article.

Conflicts of Interest: All authors declare that there are no personal or commercial relationships related to this work that would lead to a conflict of interest.

References

1. Nolte, D.; Tsang, C.K.; Zhang, K.Y.; Ding, Z.; Kedgley, A.E.; Bull, A.M.J. Non-linear scaling of a musculoskeletal model of the lower limb using statistical shape models. *J. Biomech.* **2016**, *49*, 3576–3581. [CrossRef]
2. Zhang, J.; Besier, T.F. Accuracy of femur reconstruction from sparse geometric data using a statistical shape model. *Comput. Methods Biomech. Biomed. Eng.* **2017**, *20*, 566–576. [CrossRef] [PubMed]
3. Nardini, F.; Belvedere, C.; Sancisi, N.; Conconi, M.; Leardini, A.; Durante, S.; Parenti Castelli, V. An Anatomical-Based Subject-Specific Model of In-Vivo Knee Joint 3D Kinematics From Medical Imaging. *Appl. Sci.* **2020**, *10*, 2100. [CrossRef]
4. Osti, F.; Santi, G.M.; Neri, M.; Liverani, A.; Frizziero, L.; Stilli, S.; Maredi, E.; Zarantonello, P.; Gallone, G.; Stallone, S.; et al. CT Conversion Workflow for Intraoperative Usage of Bony Models: From DICOM Data to 3D Printed Models. *Appl. Sci.* **2019**, *9*, 708. [CrossRef]
5. Belvedere, C.; Siegler, S.; Fortunato, A.; Caravaggi, P.; Liverani, E.; Durante, S.; Ensini, A.; Konow, T.; Leardini, A. New comprehensive procedure for custom-made total ankle replacements: Medical imaging, joint modeling, prosthesis design, and 3D printing. *J. Orthop. Res.* **2019**, *37*, 760–768. [CrossRef]
6. Xia, R.Z.; Zhai, Z.J.; Chang, Y.Y.; Li, H.W. Clinical Applications of 3-Dimensional Printing Technology in Hip Joint. *Orthop. Surg.* **2019**, *11*, 533–544. [CrossRef]
7. Galvez, M.; Asahi, T.; Baar, A.; Carcuro, G.; Cuchacovich, N.; Fuentes, J.A.; Mardones, R.; Montoya, C.E.; Negrin, R.; Otayza, F.; et al. Use of Three-dimensional Printing in Orthopaedic Surgical Planning. *J. Am. Acad. Orthop. Surg. Glob. Res. Rev.* **2018**, *2*, e071. [CrossRef]
8. Parthasarathy, J. 3D modeling, custom implants and its future perspectives in craniofacial surgery. *Ann. Maxillofac. Surg.* **2014**, *4*, 9–18. [CrossRef]
9. Malik, H.H.; Darwood, A.R.; Shaunak, S.; Kulatilake, P.; El-Hilly, A.A.; Mulki, O.; Baskaradas, A. Three-dimensional printing in surgery: A review of current surgical applications. *J. Surg. Res.* **2015**, *199*, 512–522. [CrossRef]
10. Martelli, N.; Serrano, C.; van den Brink, H.; Pineau, J.; Prognon, P.; Borget, I.; El Batti, S. Advantages and disadvantages of 3-dimensional printing in surgery: A systematic review. *Surgery* **2016**, *159*, 1485–1500. [CrossRef]
11. Auricchio, F.; Marconi, S. 3D printing: Clinical applications in orthopaedics and traumatology. *EFORT Open Rev.* **2016**, *1*, 121–127. [CrossRef] [PubMed]
12. Belvedere, C.; Cadossi, M.; Mazzotti, A.; Giannini, S.; Leardini, A. Fluoroscopic and Gait Analyses for the Functional Performance of a Custom-Made Total Talonavicular Replacement. *J. Foot Ankle Surg.* **2017**, *56*, 836–844. [CrossRef] [PubMed]
13. Battaglia, S.; Badiali, G.; Cercenelli, L.; Bortolani, B.; Marcelli, E.; Cipriani, R.; Contedini, F.; Marchetti, C.; Tarsitano, A. Combination of CAD/CAM and Augmented Reality in Free Fibula Bone Harvest. *Plast. Reconstr. Surg. Glob. Open* **2019**, *7*, e2510. [CrossRef] [PubMed]
14. Bahraminasab, M. Challenges on optimization of 3D-printed bone scaffolds. *Biomed. Eng. Online* **2020**, *19*, 69. [CrossRef]

15. Van Eijnatten, M.; van Dijk, R.; Dobbe, J.; Streekstra, G.; Koivisto, J.; Wolff, J. CT image segmentation methods for bone used in medical additive manufacturing. *Med. Eng. Phys.* **2018**, *51*, 6–16. [CrossRef]
16. King, A.I. A review of biomechanical models. *J. Biomech. Eng.* **1984**, *106*, 97–104. [CrossRef]
17. Leardini, A.; Belvedere, C.; Nardini, F.; Sancisi, N.; Conconi, M.; Parenti-Castelli, V. Kinematic models of lower limb joints for musculo-skeletal modelling and optimization in gait analysis. *J. Biomech.* **2017**, *62*, 77–86. [CrossRef]
18. Galbusera, F.; Cina, A.; Panico, M.; Albano, D.; Messina, C. Image-based biomechanical models of the musculoskeletal system. *Eur. Radiol. Exp.* **2020**, *4*, 49. [CrossRef]
19. An, G.; Hong, L.; Zhou, X.B.; Yang, Q.; Li, M.Q.; Tang, X.Y. Accuracy and efficiency of computer-aided anatomical analysis using 3D visualization software based on semi-automated and automated segmentations. *Ann. Anat.* **2017**, *210*, 76–83. [CrossRef]
20. Bucking, T.M.; Hill, E.R.; Robertson, J.L.; Maneas, E.; Plumb, A.A.; Nikitichev, D.I. From medical imaging data to 3D printed anatomical models. *PLoS ONE* **2017**, *12*, e0178540. [CrossRef]
21. Durastanti, G.; Leardini, A.; Siegler, S.; Durante, S.; Bazzocchi, A.; Belvedere, C. Comparison of cartilage and bone morphological models of the ankle joint derived from different medical imaging technologies. *Quant. Imaging Med. Surg.* **2019**, *9*, 1368–1382. [CrossRef] [PubMed]
22. Kresanova, Z.; Kostolny, J. Comparison of Software for Medical Segmentation. *Cent. Eur. Res. J.* **2018**, *4*, 66–80.
23. Tan, C.J.; Parr, W.C.H.; Walsh, W.R.; Makara, M.; Johnson, K.A. Influence of Scan Resolution, Thresholding, and Reconstruction Algorithm on Computed Tomography-Based Kinematic Measurements. *J. Biomech. Eng.* **2017**, *139*, 104503. [CrossRef] [PubMed]
24. Huotilainen, E.; Jaanimets, R.; Valasek, J.; Marcian, P.; Salmi, M.; Tuomi, J.; Makitie, A.; Wolff, J. Inaccuracies in additive manufactured medical skull models caused by the DICOM to STL conversion process. *J. Craniomaxillofac. Surg.* **2014**, *42*, e259–e265. [CrossRef]
25. Matsiushevich, K.; Belvedere, C.; Leardini, A.; Durante, S. Quantitative comparison of freeware software for bone mesh from DICOM files. *J. Biomech.* **2019**, *84*, 247–251. [CrossRef]
26. Lee, L.; Liew, S. A survey of medical image processing tools. In Proceedings of the 4th International Conference on Software Engineering and Computer Systems (ICSECS), Kuantan, Malaysia, 27–29 June 2011; pp. 171–176.
27. Argüello, D.; Sánchez Acevedo, H.G.; González-Estrada, O.A. Comparison of segmentation tools for structural analysis of bone tissues by finite elements. *J. Phys.* **2019**, *1386*, 012113. [CrossRef]
28. Virzi, A.; Muller, C.O.; Marret, J.B.; Mille, E.; Berteloot, L.; Grevent, D.; Boddaert, N.; Gori, P.; Sarnacki, S.; Bloch, I. Comprehensive Review of 3D Segmentation Software Tools for MRI Usable for Pelvic Surgery Planning. *J. Digit. Imaging* **2020**, *33*, 99–110. [CrossRef]
29. Fourie, Z.; Damstra, J.; Schepers, R.H.; Gerrits, P.O.; Ren, Y. Segmentation process significantly influences the accuracy of 3D surface models derived from cone beam computed tomography. *Eur. J. Radiol.* **2012**, *81*, e524–e530. [CrossRef]
30. Kamio, T.; Suzuki, M.; Asaumi, R.; Kawai, T. DICOM segmentation and STL creation for 3D printing: A process and software package comparison for osseous anatomy. *3D Print Med.* **2020**, *6*, 17. [CrossRef]
31. Ahn, C.; Bui, T.D.; Lee, Y.W.; Shin, J.; Park, H. Fully automated, level set-based segmentation for knee MRIs using an adaptive force function and template: Data from the osteoarthritis initiative. *Biomed. Eng. Online* **2016**, *15*, 99. [CrossRef]
32. Huang, J.; Jian, F.; Wu, H.; Li, H. An improved level set method for vertebra CT image segmentation. *Biomed. Eng. Online* **2013**, *12*, 48. [CrossRef] [PubMed]
33. Sander, I.M.; McGoldrick, M.T.; Helms, M.N.; Betts, A.; van Avermaete, A.; Owers, E.; Doney, E.; Liepert, T.; Niebur, G.; Liepert, D.; et al. Three-dimensional printing of X-ray computed tomography datasets with multiple materials using open-source data processing. *Anat. Sci. Educ.* **2017**, *10*, 383–391. [CrossRef] [PubMed]
34. Becker, K.; Lipprandt, M.; Röhrig, R.; Neumuth, T. Digital health—Software as a medical device in focus of the medical device regulation (MDR). *IT Inf. Technol.* **2019**, *61*, 211–218. [CrossRef]
35. Wallner, J.; Schwaiger, M.; Hochegger, K.; Gsaxner, C.; Zemann, W.; Egger, J. A review on multiplatform evaluations of semi-automatic open-source based image segmentation for cranio-maxillofacial surgery. *Comput. Methods Programs Biomed.* **2019**, *182*, 105102. [CrossRef]
36. Soodmand, E.; Kluess, D.; Varady, P.A.; Cichon, R.; Schwarze, M.; Gehweiler, D.; Niemeyer, F.; Pahr, D.; Woiczinski, M. Interlaboratory comparison of femur surface reconstruction from CT data compared to reference optical 3D scan. *Biomed. Eng. Online* **2018**, *17*, 29. [CrossRef]
37. Ortolani, M.; Leardini, A.; Pavani, C.; Scicolone, S.; Girolami, M.; Bevoni, R.; Lullini, G.; Durante, S.; Berti, L.; Belvedere, C. Angular and linear measurements of adult flexible flatfoot via weight-bearing CT scans and 3D bone reconstruction tools. *Sci. Rep.* **2021**, *11*, 16139. [CrossRef]
38. De Carvalho, K.A.M.; Walt, J.S.; Ehret, A.; Tazegul, T.E.; Dibbern, K.; Mansur, N.S.B.; Lalevee, M.; de Cesar Netto, C. Comparison between Weightbearing-CT semiautomatic and manual measurements in Hallux Valgus. *Foot Ankle Surg.* **2022**, *28*, 518–525. [CrossRef]

applied sciences

MDPI

Article

LiverNet: Diagnosis of Liver Tumors in Human CT Images

Khaled Alawneh [1], Hiam Alquran [2,3], Mohammed Alsalatie [4], Wan Azani Mustafa [5,6,*], Yazan Al-Issa [7], Amin Alqudah [7] and Alaa Badarneh [2]

1 Department of Radiology, Faculty of Medicine, Jordan University of Science and Technology, Irbid 22110, Jordan; kzalawneh0@just.edu.jo
2 Department of Biomedical Systems and Informatics Engineering, Yarmouk University, Irbid 21163, Jordan; heyam.q@yu.edu.jo (H.A.); alaa_aaa@yu.edu.jo (A.B.)
3 Department of Biomedical Engineering, Jordan University of Science and Technology, Irbid 22110, Jordan
4 The Institute of Biomedical Technology, King Hussein Medical Center, Royal Jordanian Medical Service, Amman 11855, Jordan; mhmdsliti312@gmail.com
5 Faculty of Electrical Engineering Technology, Campus Pauh Putra, University Malaysia Perlis, Arau 02000, Malaysia
6 Advanced Computing, Centre of Excellence (CoE), University Malaysia Perlis, Arau 02000, Malaysia
7 Department of Computer Engineering, Yarmouk University, Irbid 22110, Jordan; alissay@yu.edu.jo (Y.A.-I.); amin.alqudah@yu.edu.jo (A.A.)
* Correspondence: wanazani@unimap.edu.my

Abstract: Liver cancer contributes to the increasing mortality rate in the world. Therefore, early detection may lead to a decrease in morbidity and increase the chance of survival rate. This research offers a computer-aided diagnosis system, which uses computed tomography scans to categorize hepatic tumors as benign or malignant. The 3D segmented liver from the LiTS17 dataset is passed through a Convolutional Neural Network (CNN) to detect and classify the existing tumors as benign or malignant. In this work, we propose a novel light CNN with eight layers and just one conventional layer to classify the segmented liver. This proposed model is utilized in two different tracks; the first track uses deep learning classification and achieves a 95.6% accuracy. Meanwhile, the second track uses the automatically extracted features together with a Support Vector Machine (SVM) classifier and achieves 100% accuracy. The proposed network is light, fast, reliable, and accurate. It can be exploited by an oncological specialist, which will make the diagnosis a simple task. Furthermore, the proposed network achieves high accuracy without the curation of images, which will reduce time and cost.

Keywords: computed tomography; hepatic tissue; ResNet50; CAD

Citation: Alawneh, K.; Alquran, H.; Alsalatie, M.; Mustafa, W.A.; Al-Issa, Y.; Alqudah, A.; Badarneh, A. LiverNet: Diagnosis of Liver Tumors in Human CT Images. *Appl. Sci.* **2022**, *12*, 5501. https://doi.org/10.3390/app12115501

Academic Editors: Michalis Vrigkas, Christophoros Nikou and Ioannis A. Kakadiaris

Received: 18 April 2022
Accepted: 27 May 2022
Published: 29 May 2022

1. Introduction

Liver cancer (LC) is a well-known condition across the world. It is among the most frequent types of cancer that may affect humans [1]. It is a lethal disease spreading around the globe, particularly in underdeveloped nations [2]. The liver is the body's biggest internal organ. Hepatic cancer detection is difficult given the heterogeneous nature of liver tissues. The mortality rate of primary liver cancer can be reduced if it is detected earlier. For detecting the damaged region in liver images, multiple classification algorithms have been implemented [3]. The liver is both required for living and susceptible to a variety of diseases. CT examinations may be utilized to plan and deliver radiation treatment to tumors, as well as to assist biopsies and other less invasive procedures. Manual CT image segmentation and classification is a time-consuming and inefficient method, which is unfeasible for vast amounts of data. Manual interaction is not required with fully automatic and unsupervised approaches [4]. The computer-aided diagnosis of live tumors in CT images requires automatic tumor detection and segmentation. In low-contrast images, the low-level images are too faint to identify, making it a difficult process [5]. Tumor detection and segmentation are critical pre-treatment measures in the computer-aided diagnosis of liver tumors [6,7]. In the liver, there are several different forms of tumors. The visual appearance of various tumors varies, and their visual appearance varies once the contrast medium is administered. Computer-aided diagnosis might be difficult when it comes to segmenting the liver from CT scan images accurately. Automatic liver segmentation is the initial and most important stage in the diagnosing process [8,9].

Radiologists face a difficult problem in identifying and classifying liver tumors. The liver parenchyma must be separated from the abdomen, and the liver cells with the least alteration must be classified as malignant or benign tumors. Owing to its excellent cross-sectional view, outstanding spatial resolution, quick interpretation, and strong signal-to-noise ratio (SNR), CT images remain one of the top modalities of choice. Magnetic Resonance Imaging (MRI), Positron emission tomography (PET), and Ultra Sound (US) are the other major Liver-imaging modalities. CT examinations may be performed for proper planning and managing tumor treatments, including guiding biopsies and other easily established processes. For huge amounts of data, manual segmentation and Computed Axial Tomography (CAT) image categorization are demanding and time-consuming operations. Computer-aided diagnosis (CAD) systems are a type of medical imaging that acts as a second opinion for doctors when interpreting images. Upon creating the final output, the CAD systems are interactive/semi-automated and include the results of the medical practitioner. This contrasts with a fully automated system, wherein the computer software makes all choices. CAD systems have a critical role in the early diagnosis of liver disease, lowering the fatality rate from liver cancer [10]. The utilization of CT images to identify the liver disease is prevalent. Given the various intensities, it might be challenging for even competent radiologists to remark on the type, category, and level of the tumor immediately from the CT image. Designing and developing computer-assisted imaging techniques to aid physicians/doctors in enhancing their diagnoses has become increasingly significant in recent years [11]. The diagnosis and treatment strategy are determined by classifying the lesion type and time based on CT images, which demands professional knowledge and expertise to categorize. Once the workload is severe, fatigue is common, and even competent senior specialists have trouble preventing a misdiagnosis. Deep learning may overcome the limitations of conventional machine learning, for instance, the time required to retrieve image features and conduct dimensionality reduction manually, giving high-dimensional image features. It is critical to use deep learning to aid doctors in diagnosis. The poor accuracy of tumor classification, the limited capability of feature extraction, and the sparse dataset remain challenges in the current medical image classification task [12].

In 2018, Amita Das et al. [13] developed a Watershed Gaussian Deep Learning algorithm for classifying three forms of liver cancer, including hepatocellular carcinoma, hemangioma, and metastatic carcinoma, utilizing 225 images. The watershed algorithm was utilized to segregate the liver, Gaussian Mixture Models (GMM) were utilized to detect the lesion region, and retrieved characteristics were fed into a Deep Neural Network (DNN). They were able to obtain 97.72% specificity, 100% sensitivity, and 98.38% testing accuracy. Consequently, Koichiro Yasaka et al. [14] trained a Convolutional Neural Network (CNN) to distinguish liver lesions into five categories utilizing 1068 images taken in 2013 from 460 patients and enhanced them by a factor of 52. Note that three max-pooling layers, six convolutional layers, and three fully connected layers made up the CNN. They had a median accuracy of 84% and a 92% Area Under the Curve (AUC). Moreover, Kakkar et al. [15] utilized the LiTS dataset to segment the liver, utilizing the Morphological Snake method, and predicted the liver centroid utilizing an Artificial Neural Network (ANN). They obtained a 98.11% accuracy, 88% Dice Index, and 87.71% F1-score utilizing the LiTS dataset. Furthermore, Rania Ghoniem [16] employed SegNet-UNet-BCO and LeNet5-BCO combinations to segment and categorize liver lesions in 2020, combining bio-inspired concepts with deep learning models. The models were trained to utilize the Radiopaedia and LiTS datasets, and the LiTS dataset yielded a 97.6% F1-score, 98.2% specificity, 97% Dice Index, 96.4% Jaccard Index, and 98.5% accuracy. To identify liver tumors automatically, Muhammad Suhaib Aslam et al. [17] utilized the ResUNet, a hybrid UNet and ResNet framework. Relying on the publicly accessible 3D-IRCADb01 dataset, they were able to attain a 99% accuracy and a 95% F1-score. In addition, Jiarong Zhou et al. [18] presented a multi-scale and multimodal structure in 2021, utilizing a hierarchical CNN to automatically detect and categorize focal liver lesions. Following binary class discrimination, the model produced six classes. They attained an average accuracy of 82.5% in discriminating malignant and benign tumors utilizing 3D ResNet-18 and 73.4% in solving the six classes issue. Consequently, Yasmeen Al-Saeed et al. [19] presented a comprehensive framework for separating cancerous and non-cancerous lesions in 2022. The framework is divided into three phases: liver segmentation, tumor segmentation, and lesion classification with an SVM classifier. They employed a combination of textual and statistical elements to analyze the LiTS17, MICCAI-Silver07, and 3Dircadb liver datasets. The LiTS17 dataset obtained 95.57% accuracy, 96.23% sensitivity, 95.83% specificity, and 98.2% AUC, while the 3Dircadb dataset achieved 96.88% accuracy, 97.32% sensitivity, 97.65% specificity, and 98.64% AUC. Mubasher Hussain et al. [20] introduced a revolutionary, fully automated system for liver tumor classification, which employs computer vision and machine learning. A Gabor filter was employed to denoise the images, and the Correlation-based Feature Selection (CFS) approach was employed to maximize the features. On a 17 × 17 Region of Interest (ROI), they obtained 97.48% accuracy using Random Forest and 97.08% accuracy utilizing Random Trees.

According to the literature review, the subject of medical imaging is becoming more important as the demand for a precise and efficient diagnosis in a short amount of time grows. The liver serves a variety of activities, including vascular, metabolic, secretory, and excretory. CT is a medical imaging method that doctors can use to examine pathological abnormalities in the liver. The fundamental issue with liver segmentation from CT images is the poor contrast between the intensities of the liver and adjacent organs. In addition, the liver might appear in several dimensions, making identification and segmentation even more challenging [21]. The categorization of CT images is a time-consuming and difficult operation, which is impracticable when dealing with enormous amounts of data. Manual interaction is not required with fully automatic and unsupervised approaches. Our suggested study method gives an efficient liver CT scan image classification that will be useful in medical datasets, particularly in feature selection and classification. Manually detecting liver tumors is time-consuming and tiresome; however, CAD is critical in automatically recognizing liver abnormalities. In this section, we assess and review recent breakthroughs in CT-based detection of liver tumors, with a focus on deep learning techniques that leverage the LiTS dataset. In CT images, the liver is segmented from the rest of the abdomen, utilizing a 3D technique and morphological processing. The tumor is extracted from the segmented liver area using CNN. A lot of research studies have been done to categorize liver tumor disease. Patients diagnosed with a liver tumor early on will have a better chance of being treated quickly [22]. The remainder of the article is arranged as follows: Section 2 outlines the suggested method's technique. The experimental results and comparisons with a few selected approaches are shown in Section 3, and the study is concluded in Section 4.

2. Materials and Methods

The method that has been utilized in this paper is shown in Figure 1. The process starts from the segmented CT liver volume, which is resized to be compatible with the input layer of the proposed CNN and the existing ResNet50. The features are extracted automatically and then passed to a support vector machine classifier to discriminate between two classes of benign and malignant liver tumors.

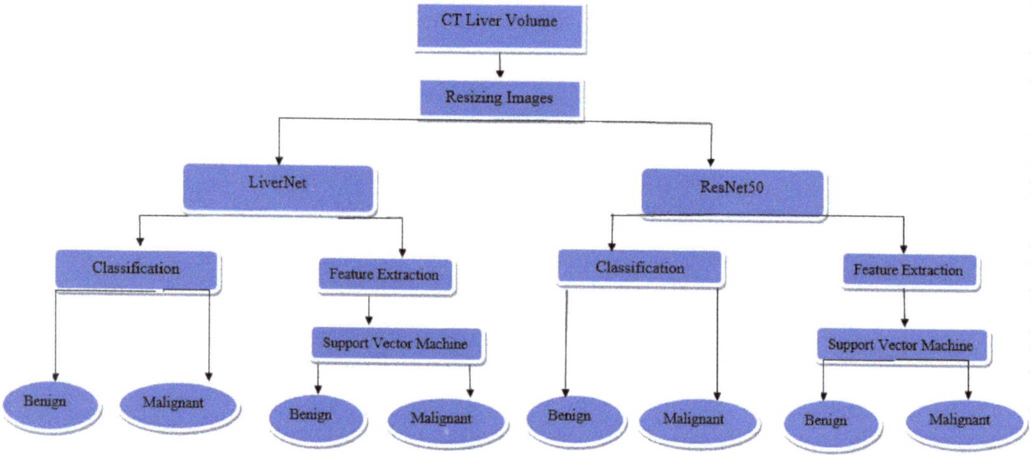

Figure 1. The proposed method.

2.1. Dataset

The dataset was created as a consequence of liver tumor segmentation, which was held in connection with the IEEE International Symposium on Biomedical Imaging (ISBI) 2017 and the International Conference on Medical Image Computing and Computer-Assisted Intervention (MICCAI) 2017. The liver images from a 3D CT image have been segmented and released. In the axial direction, pixel sizes range from 0.56 mm to 1.0 mm, and in the z-direction, they range from 0.45 mm to 6.0 mm. The number of slices per CT scan varies from 42 to 1026, and all slices were resized to 150 pixels in size. The segmented data are published in [23] and are not labeled. The data are diagnosed by the radiologist, and the following table describes the label for each case. Table 1 shows the diagnosis of the human radiologist for each volume.

Table 1. The diagnosis of liver CT volumes by Radiologist Diagnosis.

Volume#	Radiologist Label	Volume#	Radiologist Label	Volume#	Radiologist Label	Volume#	Radiologist Label	Volume#	Radiologist Label
1	malignant	27	malignant	53	malignant	79	malignant	105	benign
2	malignant	28	malignant	54	Normal	80	malignant	106	benign
3	malignant	29	malignant	55	malignant	81	malignant	107	malignant
4	malignant	30	malignant	56	malignant	82	benign	108	malignant
5	malignant	31	malignant	57	benign	83	malignant	109	malignant
6	benign	32	malignant	58	benign	84	malignant	110	benign
7	malignant	33	Normal	59	malignant	85	benign	111	malignant
8	malignant	34	malignant	60	benign	86	benign	112	benign
9	malignant	35	benign	61	malignant	87	malignant	113	benign
10	malignant	36	malignant	62	benign	88	benign	114	malignant
11	malignant	37	benign	63	benign	89	malignant	115	malignant
12	malignant	38	malignant	64	benign	90	benign	116	malignant
13	benign	39	Normal	65	malignant	91	benign	117	benign
14	malignant	40	malignant	66	benign	92	malignant	118	malignant
15	malignant	41	malignant	67	malignant	93	malignant	119	malignant
16	benign	42	benign	68	malignant	94	malignant	120	malignant
17	malignant	43	benign	69	malignant	95	malignant	121	benign
18	malignant	44	benign	70	malignant	96	malignant	122	benign
19	malignant	45	benign	71	malignant	97	malignant	123	benign
20	malignant	46	malignant	72	benign	98	benign	124	malignant
21	malignant	47	malignant	73	benign	99	malignant	125	malignant
22	malignant	48	benign	74	malignant	100	malignant	126	malignant
23	malignant	49	malignant	75	malignant	101	malignant	127	benign
24	malignant	50	malignant	76	malignant	102	malignant	128	benign
25	benign	51	benign	77	benign	103	malignant	129	malignant
26	Normal	52	malignant	78	malignant	104	benign	130	malignant

The number of benign cases is 39, the number of malignant cases is 85, whereas 6 cases are diagnosed with no lesions, which means they are normal. The normal cases are excluded from the dataset because they are not sufficient for classification. The proposed method is just designed based on benign and malignant cases.

Figure 2 describes the malignant liver slice, the segmented liver, and the 3D view of the liver. On the other hand, Figure 3 represents the benign case of the liver and its corresponding segment with its 3D view.

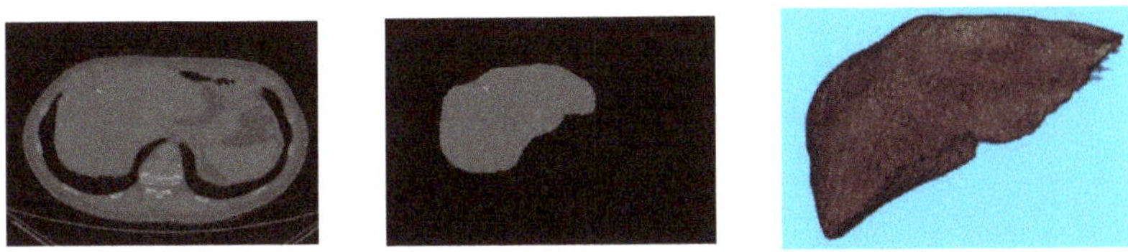

(**a**) (**b**) (**c**)

Figure 2. (**a**) liver slice; (**b**) segmented liver; (**c**) 3D view of malignant liver.

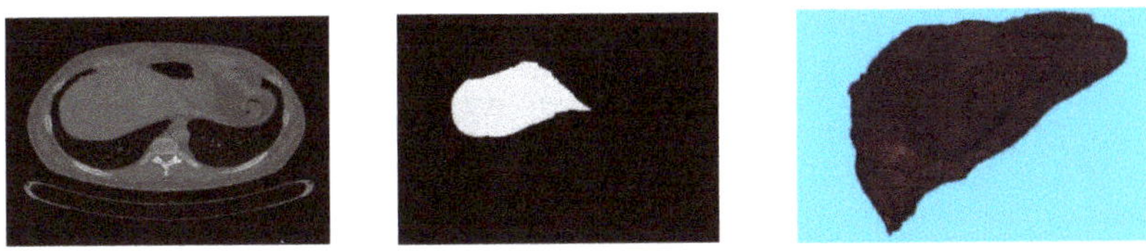

(**a**) (**b**) (**c**)

Figure 3. (**a**) liver slice; (**b**) segmented liver; (**c**) 3D view of benign liver.

The benign segmented liver is augmented with a scale [0.9–1] rotated [2°–5°]. The data are also translated in the x direction with [1–2], y direction of [1–1.5], and z direction of [0.9–1.2]. The resultant beginning images after augmentation are 75 images. Table 2 shows the number of images before and after augmentation.

Table 2. The number of images after and before augmentation.

	Benign	**Malignant**	**Total**
Number of volume images before augmentation	39	85	124
Number of volume images after augmentation	78	85	**163**

2.2. *Deep Learning*

It is known that deep learning models need large data sets to train. Many scholars have used transfer learning to tune a pre-trained model to perform a certain task to overcome this issue. In this work, two pre-trained neural networks have been used, namely ResNet50 and Resnet101 [24–28], as shown in Figure 4, respectively. The ImageNet dataset was utilized for training these two architectures. ResNet is a deep convolutional neural network model with shortcut connections that bypass one or more layers. The number of output feature maps in this type of network is similar to the number of filters in the layer. The number of filters doubles as the size of the feature map is lowered. Down sampling is performed in a convolution layer with a stride of two, and then batch normalization and the ReLU function are applied.

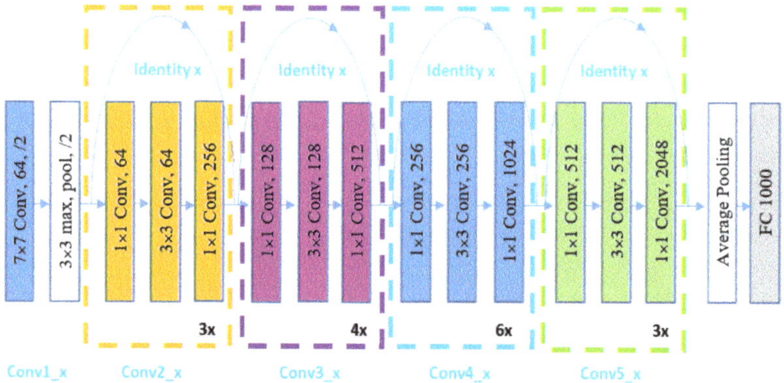

Figure 4. ResNet50 General Structure.

Further details of both networks' architectures are explained in Figure 4. The first convolutional layer in both networks will output a feature map of size 112 × 112 × 64 after applying 64 distinct filters of size 7 × 7 × 3 over the input of size 224 × 224. The input feature map is then processed, utilizing a max-pooling layer with a filter of 3 × 3, resulting in a feature map of 56 × 56 × 64. Furthermore, the second convolutional layer contains three building blocks, where each block contains three convolutional layers. As a result, there are nine sub-convolutional layers in the second convolutional layer. The third convolutional layer is made up of four blocks, each of which has three sub-convolutional layers. Thus, there are 12 sub-convolutional layers in the third convolutional layer. In terms of the fourth convolutional layer, ResNet50 comprises six blocks.

LiverNet

The proposed LiverNet model is light, and consists of eight layers. It consists of an input layer with size 223 × 223 × 147 × 1, a 3D convolutional layer with kernel size of 5 × 5 × 5 with six filters, and a stride by two. The output of the first layer is inserted into a 3D average pool layer of size 2 × 2 × 2, along with stride by two. This layer plays a crucial role in decreasing data variances and maintaining the most critical elements. Finally, the ReLU activation function receives the output from the previous layers, and the active output is sent into a 10-neuron fully connected layer. Afterwards, the result is sent to a fully connected layer with two neurons equivalent to the number of planned classes. The suggested network flow chart is illustrated in Figure 5.

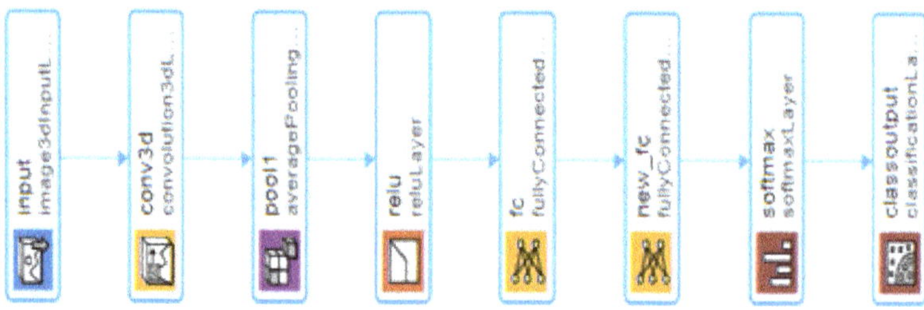

Figure 5. The structure of the proposed network.

The fully connected layer is usually terminated with a softmax layer, which implements a softmax function to its input and whose equation corresponds to the equitation [29]:

$$f(x_i) = \frac{\exp(x_i)}{\sum_j \exp(x_j)}, \tag{1}$$

in which x denotes the input vector of size K, j = 1: K, and x_i resembles the ith individual input. The Softmax function defines a range of values for the output, allowing it to be read as a probability. It is frequently employed in multivariant classifications. Moreover, the softmax layers are responsible for computing the probability of each class, whereas the classification layer is in charge of obtaining the classification results. Next, the proposed network is built using MATLAB® 2021b, and it is trained and tested using a PC with CPU Core i5-11 GEN processor, 8 GB RAM, and 1000 GB total storage. Table 3 shows the layer's information for the suggested CNN architecture.

Table 3. Layers information for the proposed LiverNet.

Layer	Information
Input Layer	Size [223 × 223 × 147]
Conv_1	Number of Filters 6 Kernel size 5 × 5 × 5 Stride 2 × 2 × 2 Padding 0
Pooling Layer	Type Average Pooling Kernel size 2 × 2 × 2 Stride 2 × 2 × 2 Padding 0
Activation Layer	ReLU
Fully-connected Layer	10 neurons
Fully-connected Layer	2 neurons
Softmax Layer	
Classification Layer	

2.3. *Classification*

The classification is performed in this article by two tracks; the first one is deep learning, and the other one is a hybrid system. The deep learning approach is utilized by passing the resize images to the pretrained ResNet50 using transfer learning to discriminate between benign and malignant classes. On the other hand, the proposed network is exploited as well for classifying the available images into malignant and benign.

The hybrid approach is utilized in this paper by using the deep learning structures as feature extractors instead of applying various image processing techniques to extract the features manually. This approach is applied two times; the first one uses ResNet50, and the other one uses the proposed Liver Network. The two extracted features from the last fully connected layer in each network are passed to the Gaussian Support Vector Machine classifier independently. The results are compared between the hybrid system, which is built based on extracted features from a pretrained ResNet50 structure, and those constructed mainly on the extracted features from the proposed LiverNet. On the top of that, the corresponding results section clarified the differences between benign and malignant classification based on deep learning approaches.

3. Results and Discussion

The data are divided into 70% training data. Meanwhile, the rest resembles testing. The transfer learning strategy is employed here to be suitable for two classes. The maximum accuracy obtained using ResNet50 is 83.7%. After taking 123 min in the training stages, Figure 6 illustrates the confusion matrix for ResNet50.

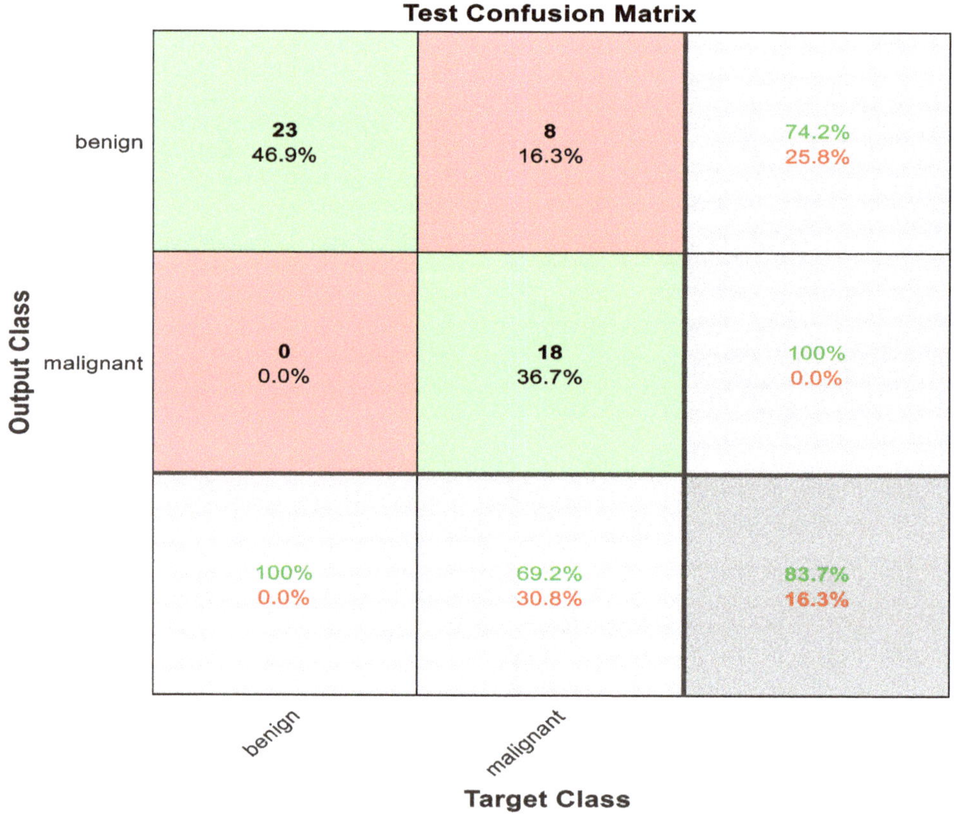

Figure 6. Test Confusion matrix of ResNet50.

The number and percentage of correct classifications by the pre-trained network are shown in the first two diagonal cells of Figure 6. For instance, 23 occurrences were categorized as benign appropriately. This represents 46.9% of the total of 49 occurrences. In the same way, 18 occurrences were accurately labeled as malignant, in which 36.7% of all cases fell into this category.

Eight of the malignant cases were misclassified as benign, accounting for 16.3% of the total 46 cases in the study. Likewise, 0 benign biopsies were wrongly labeled as malignant, accounting for 0% of the total data.

From 31 benign predictions, 74.2% were found to be correct, meanwhile 25.8% were revealed to be wrong. Of 18 malignant predictions, 100% were correct, and 0% were wrong. Of 23 benign cases, 100% were revealed to be correctly predicted as benign, while 0% were predicted as malignant. Of 26 malignant cases, 69.2% were correctly classified as malignant, while 30.8% were categorized as benign. In total, 83.7% of the predictions were revealed to be correct, while 16.3% of them were wrong. The pertained network is very badly sensitive to malignant cases. Almost 31% of the malignant cases were diagnosed as benign, which is not acceptable in medical field applications. Figure 7 shows the ROC curve for this case.

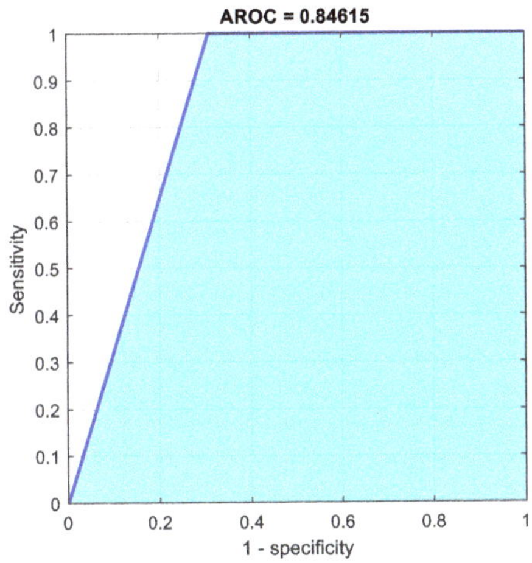

Figure 7. AROC curve using ResNet50.

The proposed net obtained a high accuracy when compared with ResNet50, and Figure 8 shows the confusion matrix of the LiverNet. Here, the accuracy reached 95.9%.

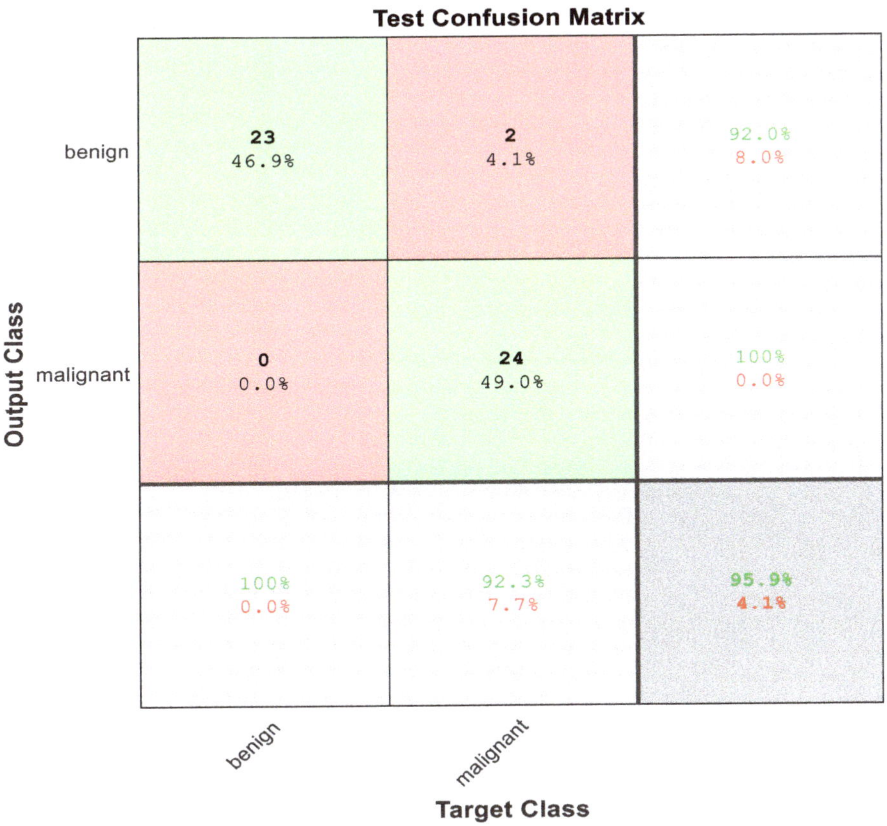

Figure 8. Test Confusion matrix of LiverNet.

The number and percentage of correct classifications by the suggested network are shown in the first two diagonal cells of Figure 7. A total of 23 occurrences, for example, were accurately categorized as benign. This represents 46.9% of the total of 49 occurrences. In the same approach, 24 occurrences were accurately labeled as malignant. This was the case in 49% of all occurrences.

Two of the malignant instances were mistakenly categorized as benign, accounting for 4.1% of the total 49 cases in the study.

From 23 benign predictions, it was revealed that 100% were correct. Meanwhile, from 24 malignant predictions, 100% were found to be correct. Moreover, from 23 benign cases, 100% were correctly predicted as benign, meanwhile, from 24 malignant cases, 92.3% were correctly classified as malignant and were discovered to be 7.7% wrong. In total, 95.9% of the predictions were found to be correct, meanwhile 4.1% were shown to be wrong. The proposed network performance was better than the pre-trained CNN. Figure 9 illustrates the ROC curve of classification using LiverNet.

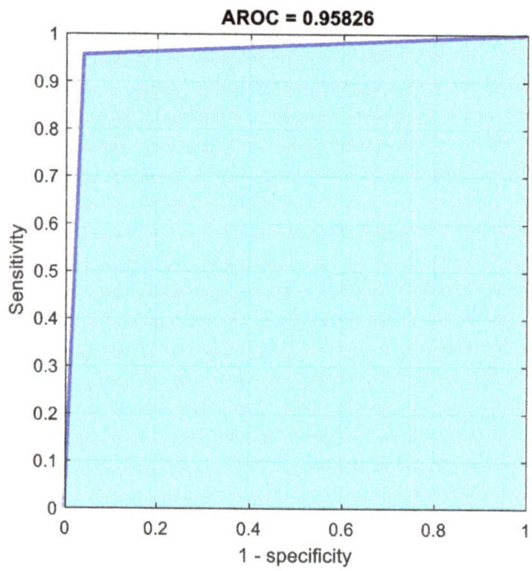

Figure 9. AROC curve using LiverNet.

The number of convolutional layers in LiverNet is one, which makes it fast in training and testing. Table 4 shows the time required for training and test phases for both the existing CNN and the proposed one.

Table 4. Comparison for training and testing time for both ResNet50 and LiverNet.

Net Work	Train	Test
ResNet50	123 min	32 s
Proposed model	89 min	22 s

In the next stage ResNet50 and LiverNet are employed as feature extractors In both networks, the two features are retrieved from the final fully connected layer. Finally, the labeled data are classified using gaussian SVM.

The model is built utilizing a Gaussian SVM classifier to distinguish between malignant and benign tumors. Figure 10 describes the confusion matrix of the Gaussian SVM using 3D graphical features of ResNet50. Here, the total accuracy reached 97%.

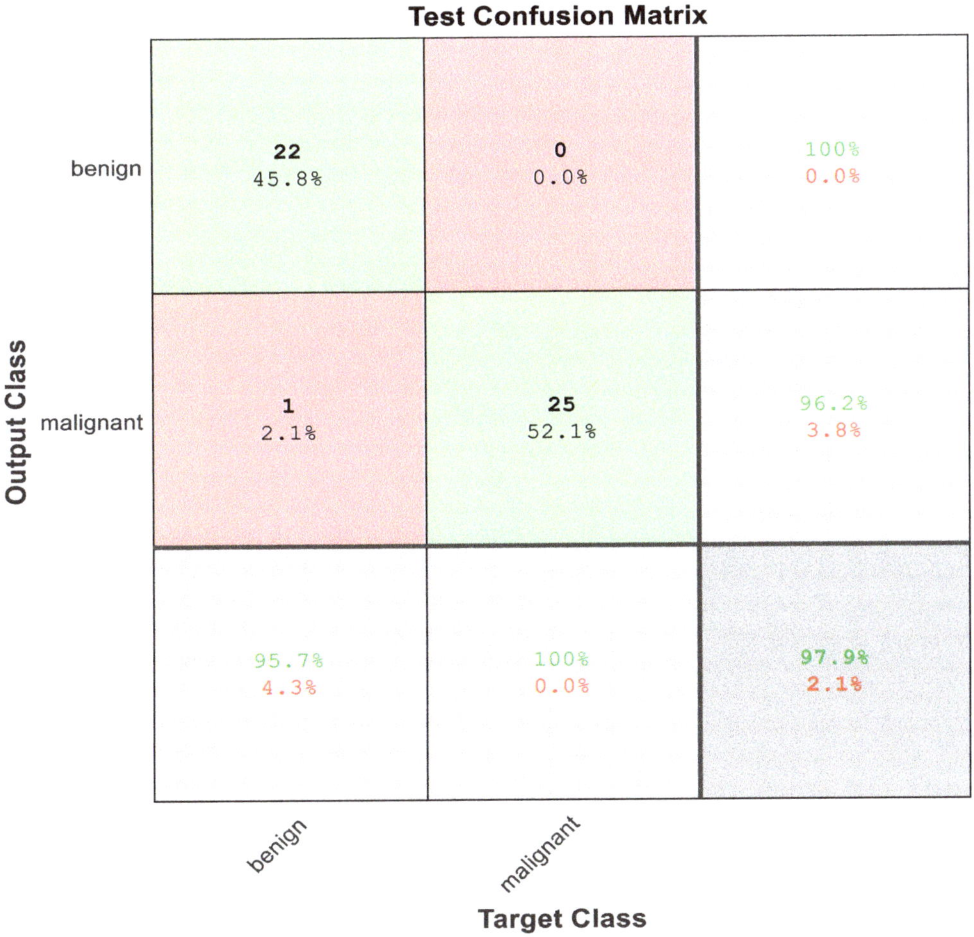

Figure 10. Test Confusion matrix of SVM with features from ResNet50.

The first two diagonal cells in Figure 7 reflect the number and percentage of correct classifications in ResNet50 utilizing a Gaussian SVM with 3D graphical features. The benign classification for the 22 cases was valid, representing 45.8% of the total 48 cases. In the same approach, 25 cases were accurately identified as malignant, which was 52.1% of the total number of cases.

One of the benign occurrences was mistakenly labeled as malignant, accounting for 2.1% of the total 48 cases in the study. From 22 benign predictions, it was stated that 100% were correct; meanwhile, from 26 malignant predictions, 96.2% were revealed to be correct. Furthermore, from 23 benign cases, 95.7% of them were correctly predicted as benign, meanwhile, from 25 malignant cases, 100% were correctly categorized as malignant. In total, 97.9% of the predictions were revealed to be correct, while 2.1% of them were wrong. Figure 11 represents the ROC curve of the hybrid system using a pretrained CNN.

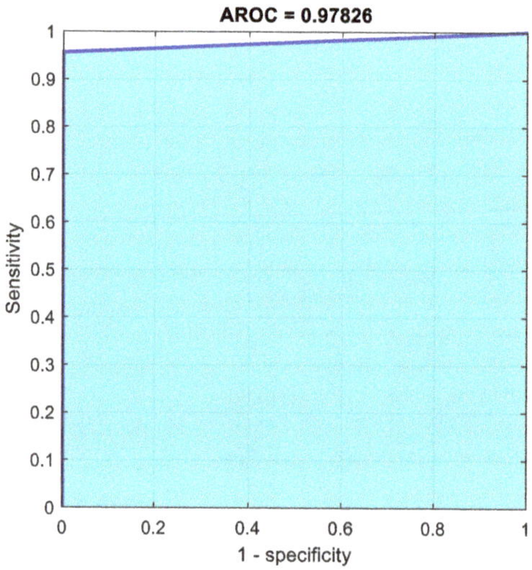

Figure 11. AROC curve of SVM with features from ResNet50.

The first two diagonal cells in Figure 12 reflect the number and percentage of correct classifications utilizing the suggested net's gaussian SVM with 3D graphical features. For example, the benign classification for the 23 occurrences was correct, representing 47.9% of the total 48 occurrences. In the same approach, 25 occurrences were accurately identified as malignant, which was 52.1% of the total number of cases.

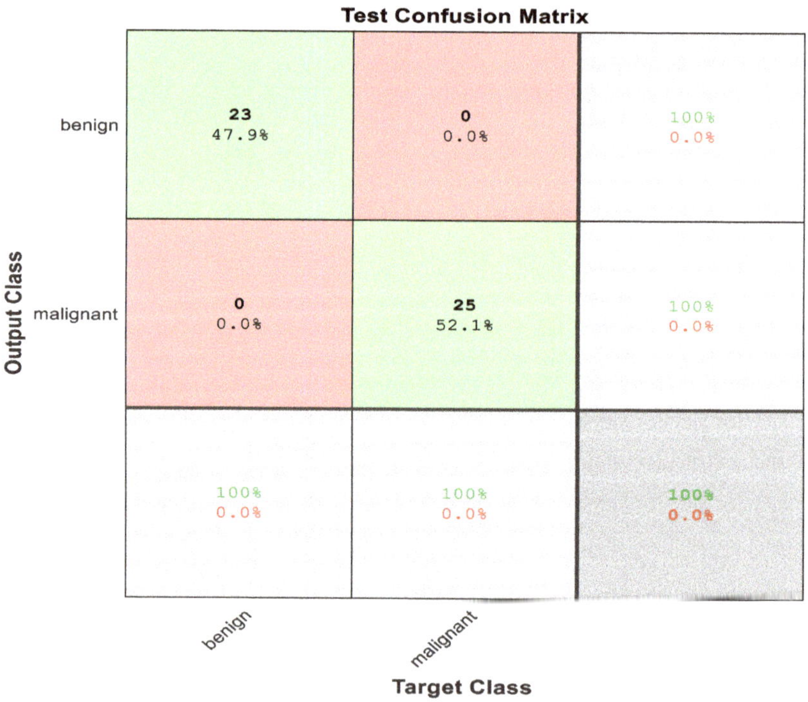

Figure 12. Test Confusion matrix of SVM with features from LiverNet.

Of 23 benign cases, 100% of them were correctly predicted as benign, meanwhile, from 25 malignant cases, 100% were found to be correctly categorized as malignant. Overall, 100% of the predictions were correct. Figure 13 shows the ROC curve of the hybrid system using LiverNet features.

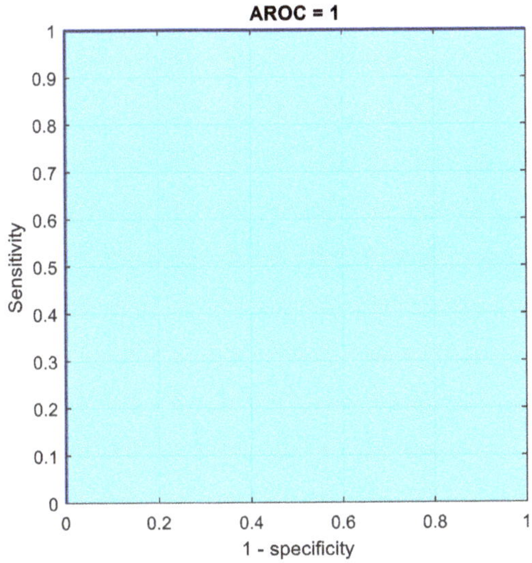

Figure 13. AROC of hybrid system with features from LiverNet.

The evaluation criteria that have been used in this paper are clear in the corresponding equations [30]. Table 5 describes the results in the deep learning track and hybrid track for both ResNet50 and the proposed LiverNet. The following equations are used to calculate the performance of the classifier [30].

$$\text{Sensitivity} = \frac{\text{TP}}{\text{TP} + \text{FN}} \tag{2}$$

$$\text{Precision} = \frac{\text{TP}}{\text{TP} + \text{FP}} \tag{3}$$

$$\text{Specificity} = \frac{\text{TN}}{\text{TN} + \text{FP}} \tag{4}$$

$$\text{Accuracy} = \frac{\text{TP}}{\text{TP} + \text{TN} + \text{FP} + \text{FN}} \tag{5}$$

where TP is True Positive, TN is True Negative, FP is False Positive, and FN is False Negative.

Table 5. Comparison between deep learning and the proposed hybrid model.

Method		Sensitivity	Precision	Specificity	Accuracy
Deep Learning	ResNet50	100	74.2	69.2	83.7
	LiverNet	100	92	92.3	95.9
Hybrid Model	ResNet50	95.7	100	100	97.9
	LiverNet	100	100	100	100

The high performance of the proposed net is clear as a feature extractor. Furthermore, Figure 14 below illustrates the high performance of the proposed approach in obtaining an accurate diagnosis of liver tumors.

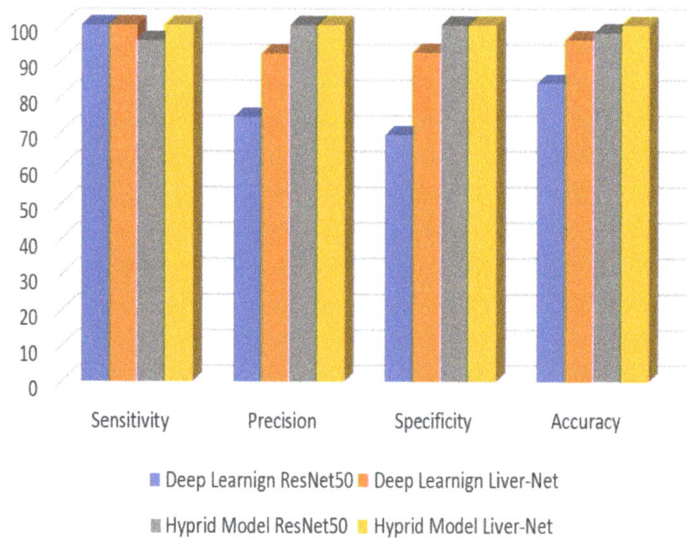

Figure 14. Accuracy, Precision, Specificity, and Sensitivity of different approaches.

The proposed method is compared with literature that has used the LiTs17 dataset. The performance of the approach achieved the highest amongst all. Table 6 shows the comparison between this study and literature with regards to the area under the curve (AUC), specificity, sensitivity, and accuracy.

Table 6. Comparison of the current study with the state of the art.

	Accuracy	Sensitivity	Specificity	AROC
Kakkar et al. [15]	98.11	-	-	-
Rania Ghoniem [16]	98.5	-	98.2	-
Yasmeen Al-Saeed et al. [19]	95.5	96.23	95.83	0.98
The Proposed Model	**100**	**100**	**100**	**1**

This paper shows the high level of confidence obtained using LiverNet as an automated feature extractor besides utilizing the benefits of machine learning to discriminate between benign and malignant liver tumors.

4. Conclusions

Patients with liver cancer have a high mortality rate attributed to the late detection of the disease. Computer-aided diagnosis systems based on a variety of medical imaging techniques can help recognize liver cancer at an early stage. With the help of both conventional machine learning and deep learning classifiers, a variety of methods have been employed to identify liver cancer. The findings of this study suggest that using CNN to automatically extract features together with SVM classifier greatly improves classification performance. Furthermore, the findings suggest that employing our suggested hybrid model can greatly reduce the processing time, which is 22 s, when contrasted to ResNet50, which takes 32 s. All performance metrics accuracy, specificity, precision, and sensitivity reached 100%. Our approach can accurately and effectively recognize tumors, even in low-contrast CT images with respect to all quantitative assessments. Lastly, we can draw the following conclusions: (1) Deep learning model performance is extremely intriguing for use in medical equipment; the experimental result demonstrates significant improvement. Moreover, the suggested technique is unaffected by discrepancies in texture and intensity across demographics, imaging devices, patients, and settings; (2) the classifier distinguishes the tumor with comparatively high precision; (3) segmentation of very small tumors is incredibly challenging, with the system being hyper-sensitive to contemplating local noise artifacts as potential tumors.

The lack of large publicly available datasets forces CAD systems to use the available small private datasets generated from hospitals and scanning facilities. This implies that additional datasets should be made available for research and classification purposes. In the future, this work can be further extended using a large clinical dataset besides applying image processing techniques to enhance the visualization of images. Using a huge dataset, a reliable and trusted system can be built and employed in clinics.

Author Contributions: Conceptualization, K.A., H.A. and M.A.; methodology, A.A., H.A., Y.A.-I. and M.A.; software, H.A., M.A. and W.A.M.; validation, K.A., H.A., Y.A.-I., W.A.M. and M.A.; formal analysis, Y.A.-I., H.A., A.B. and M.A.; writing—original draft preparation, H.A., Y.A.-I., A.A., A.B. and W.A.M.; writing—review and editing, K.A., H.A., W.A.M., M.A. and Y.A.-I.; visualization, H.A.; supervision, H.A. and W.A.M.; project administration, H.A. and K.A. All authors have read and agreed to the published version of the manuscript.

Funding: This research received no external funding.

Data Availability Statement: The dataset analyzed during the current study was derived from the LiTS (Liver Tumor Segmentation Challenge (LiTS17)) organized in conjunction with ISBI 2017 and MICCAI 2017. Available online: https://www.kaggle.com/datasets/andrewmvd/liver-tumor-segmentation (accessed on 15 January 2022).

Acknowledgments: The authors would thank the authors of the dataset for making it available online. Furthermore, they would like to thank the anonymous reviewers for their contribution towards enhancing this paper.

Conflicts of Interest: The authors declare no conflict of interest.

References

1. Al Sadeque, Z.; Khan, T.I.; Hossain, Q.D.; Turaba, M.Y. Automated detection and classification of liver cancer from CT Images using HOG-SVM model. In Proceedings of the 2019 5th International Conference on Advances in Electrical Engineering (ICAEE 2019), Dhaka, Bangladesh, 26–28 September 2019; pp. 21–26.
2. Ba Alawi, A.E.; Saeed, A.Y.A.; Radman, B.M.N.; Alzekri, B.T. A Comparative Study on Liver Tumor Detection Using CT Images. In *Lecture Notes on Data Engineering and Communications Technologies*; Springer: Cham, Switzerland, 2021; Volume 72, pp. 129–137.
3. Shanila, N.; Vinod Kumar, R.S.; Abin, N.A. Feature extraction and performance evaluation of classification algorithms for liver tumor diagnosis of abdominal computed tomography images. *J. Adv. Res. Dyn. Control Syst.* **2020**, *12*, 82–90. [CrossRef]
4. Selvathi, D.; Malini, C.; Shanmugavalli, P. Automatic segmentation and classification of liver tumor in CT images using adaptive hybrid technique and Contourlet based ELM classifier. In Proceedings of the 2013 International Conference on Recent Trends in Information Technology (ICRTIT 2013), Dubai, United Arab Emirates, 11–12 December 2013; pp. 250–256.
5. Masuda, Y.; Tateyama, T.; Xiong, W.; Zhou, J.; Wakamiya, M.; Kanasaki, S.; Furukawa, A.; Chen, Y.W. Liver tumor detection in CT images by adaptive contrast enhancement and the EM/MPM algorithm. In Proceedings of the International Conference on Image Processing (ICIP), Brussels, Belguim, 11–14 September 2011; pp. 1421–1424.
6. Hasegawa, R.; Iwamoto, Y.; Han, X.; Lin, L.; Hu, H.; Cai, X.; Chen, Y.W. Automatic Detection and Segmentation of Liver Tumors in Multi- phase CT Images by Phase Attention Mask R-CNN. In Proceedings of the Digest of Technical Papers—IEEE International Conference on Consumer Electronics, Penghu, Taiwan, 15–17 September 2021.
7. Salman, O.S.; Klein, R. Automatic Detection and Segmentation of Liver Tumors in Computed Tomography Images: Methods and Limitations. In *Intelligent Computing*; Lecture Notes in Networks and Systems; Springer: Cham, Switzerland, 2021; Volume 285, pp. 17–35.
8. Das, A.; Panda, S.S.; Sabut, S. Detection of liver tumor in CT images using watershed and hidden markov random field expectation maximization algorithm. In *Computational Intelligence, Communications, and Business Analytics*; Springer: Singapore, 2017; Volume 776, pp. 411–419.
9. Todoroki, Y.; Han, X.H.; Iwamoto, Y.; Lin, L.; Hu, H.; Chen, Y.W. Detection of liver tumor candidates from CT images using deep convolutional neural networks. In *International Conference on Innovation in Medicine and Healthcare*; Springer: Cham, Switzerland, 2018; Volume 71, pp. 140–145.
10. Devi, R.M.; Seenivasagam, V. Automatic segmentation and classification of liver tumor from CT image using feature difference and SVM based classifier-soft computing technique. *Soft Comput.* **2020**, *24*, 18591–18598. [CrossRef]
11. Krishan, A.; Mittal, D. Ensembled liver cancer detection and classification using CT images. *Proc. Inst. Mech. Eng. Part H J. Eng. Med.* **2021**, *235*, 232–244. [CrossRef] [PubMed]

12. Mao, J.; Song, Y.; Liu, Z. CT image classification of liver tumors based on multi-scale and deep feature extraction. *J. Image Graph.* **2021**, *26*, 1704–1715. [CrossRef]
13. Das, A.; Acharya, U.R.; Panda, S.S.; Sabut, S. Deep learning based liver cancer detection using watershed transform and Gaussian mixture model techniques. *Cogn. Syst. Res.* **2019**, *54*, 165–175. [CrossRef]
14. Yasaka, K.; Akai, H.; Abe, O.; Kiryu, S. Deep learning with convolutional neural network for differentiation of liver masses at dynamic contrast-enhanced CT: A preliminary study. *Radiology* **2018**, *286*, 887–896. [CrossRef] [PubMed]
15. Kakkar, P.; Nagpal, S.; Nanda, N. Automatic liver segmentation in CT images using improvised techniques. In *International Conference on Smart Health*; Lecture Notes in Computer Science; Springer: Cham, Switzerland, 2018; Volume 10983 LNCS, pp. 41–52.
16. Ghoniem, R.M. A Novel Bio-Inspired Deep Learning Approach for Liver Cancer Diagnosis. *Information* **2020**, *11*, 80. [CrossRef]
17. Aslam, M.S.; Younas, M.; Sarwar, M.U.; Shah, M.A.; Khan, A.; Uddin, M.I.; Ahmad, S.; Firdausi, M.; Zaindin, M. Liver-Tumor detection using CNN ResUNet. *Comput. Mater. Contin.* **2021**, *67*, 1899–1914. [CrossRef]
18. Zhou, J.; Wang, W.; Lei, B.; Ge, W.; Huang, Y.; Zhang, L.; Yan, Y.; Zhou, D.; Ding, Y.; Wu, J.; et al. Automatic Detection and Classification of Focal Liver Lesions Based on Deep Convolutional Neural Networks: A Preliminary Study. *Front. Oncol.* **2021**, *10*, 581210. [CrossRef] [PubMed]
19. Al-Saeed, Y.; Gab-Allah, W.A.; Soliman, H.; Abulkhair, M.F.; Shalash, W.M.; Elmogy, M. Efficient Computer Aided Diagnosis System for Hepatic Tumors Using Computed Tomography Scans. *Comput. Mater. Contin.* **2022**, *71*, 4871–4894. [CrossRef]
20. Hussain, M.; Saher, N.; Qadri, S. Computer Vision Approach for Liver Tumor Classification Using CT Dataset. *Appl. Artif. Intell.* **2022**, 1–23. [CrossRef]
21. Selvathi, D.; Priyadarsini, S.; Malini, C.; Shanmugavalli, P. Performance analysis of multi resolution transforms with kernel classifiers for liver tumor detection using CT images. *Int. J. Appl. Eng. Res.* **2014**, *9*, 30935–30952.
22. Krishan, A.; Mittal, D. Effective segmentation and classification of tumor on liver MRI and CT images using multi-kernel K-means clustering. *Biomed. Technol.* **2019**, 301–313. [CrossRef] [PubMed]
23. Soler, L.; Hostettler, A.; Agnus, V.; Charnoz, A.; Fasquel, J.; Moreau, J.; Osswald, A.; Bouhadjar, M.; Marescaux, J. *3D Image Reconstruction for Comparison of Algorithm Database: A Patient Specific Anatomical and Medical Image Database*; Tech Report; IRCAD: Strasbourg, France, 2010.
24. He, K.; Zhang, X.; Ren, S.; Sun, J. Deep residual learning for image recognition. In Proceedings of the IEEE Computer Society Conference on Computer Vision and Pattern Recognition, Las Vegas, NV, USA, 27–30 June 2016; pp. 770–778.
25. Poudel, S.; Kim, Y.J.; Vo, D.M.; Lee, S.W. Colorectal Disease Classification Using Efficiently Scaled Dilation in Convolutional Neural Network. *IEEE Access* **2020**, *8*, 99227–99238. [CrossRef]
26. Ullah, W.; Ullah, A.; Haq, I.U.; Muhammad, K.; Sajjad, M.; Baik, S.W. CNN features with bi-directional LSTM for real-time anomaly detection in surveillance networks. *Multimed. Tools Appl.* **2021**, *80*, 16979–16995. [CrossRef]
27. Wu, H.; Xin, M.; Fang, W.; Hu, H.M.; Hu, Z. Multi-Level Feature Network with Multi-Loss for Person Re-Identification. *IEEE Access* **2019**, *7*, 91052–91062. [CrossRef]
28. Chen, J.; Zhou, M.; Zhang, D.; Huang, H.; Zhang, F. Quantification of water inflow in rock tunnel faces via convolutional neural network approach. *Autom. Constr.* **2021**, *123*, 103526. [CrossRef]
29. Alqudah, A.; Alqudah, A.M.; Alquran, H.; Al-zoubi, H.R.; Al-qodah, M.; Al-khassaweneh, M.A. Recognition of handwritten arabic and hindi numerals using convolutional neural networks. *Appl. Sci.* **2021**, *11*, 1573. [CrossRef]
30. Alqudah, A.M.; Alquran, H.; Abu-Qasmieh, I.; Al-Badarneh, A. Employing image processing techniques and artificial intelligence for automated eye diagnosis using digital eye fundus images. *J. Biomim. Biomater. Biomed. Eng.* **2018**, *39*, 40–56. [CrossRef]

applied sciences

MDPI

Article

Cephalometric Landmark Detection in Lateral Skull X-ray Images by Using Improved SpatialConfiguration-Net

Martin Šavc, Gašper Sedej and Božidar Potočnik *

Faculty of Electrical Engineering and Computer Science, University of Maribor, Koroška cesta 46, 2000 Maribor, Slovenia; martin.savc@um.si (M.Š.); gasper.sedej@um.si (G.S.)
* Correspondence: bozidar.potocnik@um.si; Tel.: +386-2-220-7484

Abstract: Accurate automated localization of cephalometric landmarks in skull X-ray images is the basis for planning orthodontic treatments, predicting skull growth, or diagnosing face discrepancies. Such diagnoses require as many landmarks as possible to be detected on cephalograms. Today's best methods are adapted to detect just 19 landmarks accurately in images varying not too much. This paper describes the development of the SCN-EXT convolutional neural network (CNN), which is designed to localize 72 landmarks in strongly varying images. The proposed method is based on the SpatialConfiguration-Net network, which is upgraded by adding replications of the simpler local appearance and spatial configuration components. The CNN capacity can be increased without increasing the number of free parameters simultaneously by such modification of an architecture. The successfulness of our approach was confirmed experimentally on two datasets. The SCN-EXT method was, with respect to its effectiveness, around 4% behind the state-of-the-art on the small ISBI database with 250 testing images and 19 cephalometric landmarks. On the other hand, our method surpassed the state-of-the-art on the demanding AUDAX database with 4695 highly variable testing images and 72 landmarks statistically significantly by around 3%. Increasing the CNN capacity as proposed is especially important for a small learning set and limited computer resources. Our algorithm is already utilized in orthodontic clinical practice.

Keywords: detection of cephalometric landmarks; skull X-ray images; convolutional neural networks; deep learning; SpatialConfiguration-Net architecture; AUDAX database

Citation: Šavc, M.; Sedej, G.; Potočnik, B. Cephalometric Landmark Detection in Lateral Skull X-ray Images by Using Improved SpatialConfiguration-Net. *Appl. Sci.* **2022**, *12*, 4644. https://doi.org/10.3390/app12094644

Academic Editors: Ioannis A. Kakadiaris, Michalis Vrigkas and Christophoros Nikou

Received: 14 April 2022
Accepted: 29 April 2022
Published: 5 May 2022

1. Introduction

Cephalometry has been used for many years for the diagnosis of malformations, surgical planning and evaluation, and growth studies. This discipline relies on the identification of craniofacial landmarks [1,2]. Cephalometric analysis, or cephalometrics, is the clinical application of cephalometry to the field of orthodontics. Cephalometrics has been used in orthodontic diagnosis to evaluate the pretreatment dental and facial relationship of a patient, to evaluate changes during treatment, and to assess tooth movement and facial growth at the end of treatment [3]. The first important step in cephalometric analysis is accurate detection of cephalometric landmarks on the cephalogram, i.e., an X-ray image of the craniofacial area (shortly, a skull image). In the cephalometric assessment, certain carefully defined points should be located on the radiographs, and linear and angular measurements are then made from these points [3]. Only accurate measurements and calculations represent diagnostic aids for orthodontists.

There exist lateral and frontal cephalograms. Lateral cephalograms provide a lateral view of the skull, while the frontal cephalograms present an antero-posterior view of the skull. The lateral cephalograms will be utilized in this study. Figure 1 depicts sample lateral cephalograms, captured in a natural head position, which enables the repeatability of image capture and comparison of different cephalometric analyses.

Early attempts for computerized detection of cephalometric landmarks were found around the year 2000. Several (prototype) methods for automatic landmark identification from skull X-ray images (cephalograms) have emerged, based on heuristic features and rigid rules. These methods were highly dependent on the quality of the input images, and were adapted for a small number of landmarks [1] (the number of landmarks is meant here as the number of different types of landmarks we are looking for in each image). More mature methods, as well as learning-based approaches, emerged after 2010 [4,5]. Lindner et al. [5,6] proposed an efficient detection method based on Haar-like features and random forests (RFs). An RF was trained for each landmark in order to predict the more probable position of that landmark. Each tree in the RF voted for the likely new position. The RF regression-voting mechanism was integrated into the constrained local model framework that optimized a statistical shape model and total votes over all landmark positions. This detection system was adapted for the detection of 19 cephalometric landmarks. A similar method with RF and Haar-like appearance features was proposed by Ibragimov et al. in [4,5,7]. The difference was that a matching of the appearance shape model in a target image was sought by using a game-theoretic optimization framework. The fitted model determined the optimal landmark positions.

Recently, successful methods have emerged based on convolutional neural networks (CNN) and deep learning. We expose the four best, which are comparable in effectiveness. Chen et al., in a conference article [8], proposed the CNN-based architecture that consists of the pretrained VGG-19 net as a feature extraction module, an attentive feature pyramid fusion (AFPF) module, and a prediction module. They fused features from different levels in order to obtain high-resolution and semantically enhanced features in the AFPF module. A self-attention mechanism was utilized to learn corresponding weights for the fusion for different landmarks. Finally, a combination of heat maps and offset maps was employed in the prediction module to perform a pixel-wise regression-voting. The next conference paper is from Li et al. [9], who modeled landmarks as a graph and employed two global-to-local cascaded graph convolutional networks (GCNs) to reposition the landmarks towards the target locations. The graph signals of the landmarks were built by combining local image features and graph shape features. The authors state that their method is able to exploit the structural knowledge effectively and allow rich information exchange between landmarks for accurate coordinate estimation. The first GCN estimated a global transformation of the landmarks, while the second GCN determined local offsets to adjust the landmark coordinates further. Payer et al., in a journal article [10], introduced a CNN architecture that learns to split the localization task into two simpler sub-problems, thus reducing the overall need for large training datasets. Their fully convolutional SpatialConfiguration-Net (SCN) utilized one component to obtain locally accurate but ambiguous candidate predictions, while the other component improved robustness to ambiguities by incorporating the spatial configuration of landmarks. Since our research is based on this method, we will provide details about the SCN in the next sections. Lastly, we expose the method by Song et al. [11]. The authors proposed the usage of an individual model for each landmark, where each model was trained by the ResNet50 architecture. These constructed models were applied to smaller patches extracted from the cephalometric image. The method assumed that each patch that was passed into the model must contain the landmark that was being detected by this model. To ensure this, each testing image was aligned to every training image by using a translational registration. Landmarks from the training image with the best fit after registration were considered as centers for the extracted patches. The results obtained on the database of public cephalograms with 19 landmarks were comparable to other state-of-the-art methods. However, this method does not scale well to a larger number of cephalometric landmarks and training images.

In order for cephalometric analysis to be meaningful and useful as a diagnostic tool, it is necessary to detect as many cephalometric landmarks on the cephalogram as accurately as possible. Usage of lateral cephalograms predominates today in the field of orthodontics; therefore, we also focused on this type of cephalograms in our research (similar to the related works summarized above). The identified shortcomings of early related works indicated that these methods were adapted for a small number of cephalometric landmarks and for a small number of high-quality input images. State-of-the-art methods [8–10] are practically invariant to brightness/contrast variations, or to situations during cephalograms' capture, respectively. Additionally, an addition of new landmarks that we would like to detect with these methods is relatively simple, as we only need to supplement the learning set and retrain the CNNs (and possibly add some channels). Although state-of-the-art methods have proven to be very effective in locating cephalometric landmarks, it should be noted that these methods have been validated on only 19 landmarks and on just some hundred testing images. Thus, a research question arises as to whether the CNN architectures of these methods have sufficient capacity to localize a larger number of landmarks effectively on a larger set of testing images captured with different X-ray devices. We are tackling a real-world problem from the field of orthodontics in this research; namely, we are developing a detection method as an enhancement of the state-of-the-art, which will be able to detect a large number of cephalometric landmarks (in our study 72) on highly variable testing images. It is understood by variability that testing images are of different sizes (and different spatial resolutions), and that they were captured by using different X-ray devices in different orthodontic clinics (most likely with different device settings). On the other hand, this research also solves one of the concrete problems of the industry (e.g., the AUDAX company). Virtually every orthodontic software includes a module for detecting cephalometric landmarks. A greater number of very precisely localized landmarks of course means better usability of such software. For accurate cephalometric analyses, we need to localize as many landmarks as possible, as only in this way can we diagnose discrepancies or patients' face disharmony, predict skull growth, or plan treatments.

In this study, we will adapt the architecture of the state-of-the-art SCN network in order to detect 72 cephalometric landmarks on highly variable X-ray images. The aim is, on the one hand, to increase the capacity of the CNN (i.e., the ability to learn several different transformation functions), while maintaining approximately the same number of free parameters (degrees of freedom—DoF) as the basic SCN network has. The latter is achieved by expanding the local appearance and spatial configuration components of the SCN network, and not by a raw increase of filters' sizes and numbers of channels. Maintaining DoF while increasing network capacity is important, especially for a small learning set and limited computer resources, which is often the case in healthcare. This, in turn, means a better ability to train such an NN and prevent overfitting. The effectiveness of our proposed SCN-EXT method was confirmed experimentally by detecting 72 cephalometric landmarks on a challenging private database of 4695 cephalograms.

The contribution of this research work is summarized in

1. The development of a sophisticated landmark detection algorithm, where this algorithm is built on the state-of-the-art SpatialConfiguration-Net neural network.
2. Introduction of the most effective algorithm for the detection of 72 cephalometric landmarks on the lateral skull X-ray images.
3. The first study that assesses the effectiveness of the state-of-the-art cephalometric landmark detection algorithms on a large number of landmarks and on a large number of testing images.

This article is structured as follows. A short overview of cephalometric landmarks' classification and employed evaluation databases is given in Section 2. A novel cephalometric landmark detection algorithm based on the SpatialConfiguration-Net architecture is described in detail in Section 3. Some considerations about the proposed method implementation and CNN training are clarified in Section 4. This section also introduces the evaluation metrics used in our experiments. Section 5 presents some of the results obtained on the public and private databases, followed by Section 6, which emphasizes certain aspects of our detection method. Section 7 concludes this paper briefly with some hints about future work.

2. Experimental Methods

2.1. Cephalometric Landmarks

There are two well-known classifications of cephalometric landmarks [3], namely, (1) based on the origin, we distinguish between (i) anatomic and (ii) derived or constructed cephalometric landmarks, and (2) based on the structures involved, we differentiate between (a) hard tissue and (b) soft tissue cephalometric landmarks. Anatomic landmarks represent the actual anatomic structures of the skull (e.g., nasion, point A, point B, ANS, PNS, etc.), while derived or constructed landmarks are obtained secondarily from anatomic structures in a lateral cephalogram (e.g., gnathion, anterior point of occlusion, etc.). On the other hand, the hard tissue cephalometric landmarks represent the actual hard tissue structures of the skull, such as the nasal bone, frontal bone, maxillary bone, etc., while soft tissue landmarks, as their name suggests, are located on the soft tissues (e.g., on the forehead, nose, lips, etc.) [3]. Examples of hard tissue cephalometric landmarks are nasion, temporale, sella, menton, and gonion, while examples of soft tissues landmarks are subnasale, subspinale, stomion, soft tissue pogonion, and soft tissue gnathion [3].

2.2. Evaluation Databases

Two different databases were used to evaluate the effectiveness of the detection methods in this study, namely, the ISBI public image database with 19 annotated cephalometric landmarks on each image, and the AUDAX private image database with 72 landmarks per image.

2.2.1. ISBI Public Database

Wang et al. [5] released a public database of 400 cehpalometric images, where 19 of the more common landmarks were annotated on each image. A list of all the annotated landmarks is presented in Table 1. Radiographs were collected from 400 patients ranging from 6 to 60 years old. All cephalograms were captured by the same X-ray device. Every image was annotated manually by two experienced medical doctors. A ground truth was determined as an average of the annotations of both doctors. The images have the same dimension of 1935 × 2400 pixels with 10 pixel/mm spatial resolution.

Table 1. A list of 19 cephalometric landmarks annotated in the ISBI public database. A description of the landmarks and their significance can be found in [3].

−1i—Lower incisal incisior	+1i—Upper incisal incisior	ANS—Anterior Nasal Spine	Ar—Articulare
Gn—Gnathion	Go—Gonion	Li′—Lower lip	Ls′—Upper lip
Me—Menton	N—Nasion	Or—Orbitale	Pg—Pogonion
Pg′—Point Soft Pogonion	PNS—Posterior Nasal Spine	Po—Porion	S—Sella Turcica
Sn′—Subnasale	SS—Subspinale (Point A)	SM—Supramentale (Point B)	

This database is divided into three sets. The first 300 out of 400 images are from the 2015 Automatic Cephalometric X-Ray Landmark Detection Challenge [4]. These 300 images were split into a training set (150 images) and testing set 1 (the remaining 150 images). The 2016 Automatic Cephalometric X-Ray Landmark Detection Challenge brought another 100 images to this public database. These 100 images are denoted as testing set 2. Figure 1a depicts a sample annotated image from this public database. Landmarks are actually pixels, but they are depicted as white circles in this image.

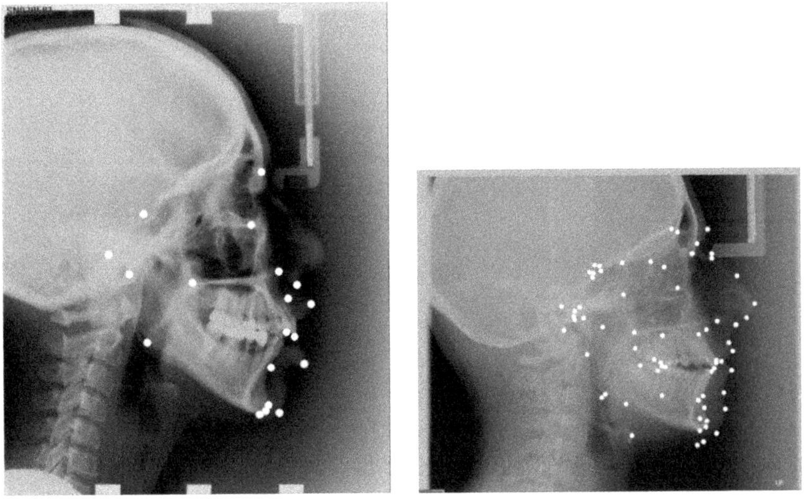

(a) (b)

Figure 1. Sample annotated cephalograms from (**a**) the ISBI public database [5], with 19 landmarks, and (**b**) the AUDAX private database, with 72 landmarks (white circles).

2.2.2. AUDAX Private Database

A private database was constructed during an industrial project between our research group and the Slovenian company AUDAX (https://audaxceph.com (accessed on 13 April 2022)), which is specialized for the development of orthodontic software. This database consists of 4695 unique skull X-ray images. We assumed that each radiograph belongs to a different subject. Information about the image spatial resolution and about the subject in the image (e.g., gender, age, health status) was not provided by AUDAX. The size of images ranged from 355 × 480 pixels (min size) to 4417 × 5963 pixels (max size). There are 287 unique image sizes in this database. On this basis, we concluded that the images were captured with just as many different X-ray devices. The five most common sizes of radiographs were as follows: 2808 × 2148 pixels (1598 images), 1000 × 900 pixels (419), 2685 × 2232 pixels (310), 1804 × 2148 pixels (309), and 1000 × 765 pixels (222). An average image size was 1740 × 2012 pixels.

Seventy-two cephalometric landmarks were annotated on each image by a single experienced orthodontist. A list of all annotated landmarks is gathered in Table 2. Most landmarks are anatomic landmarks, while the rest were constructed relative to anatomic landmarks, or were defined as intersections of particular lines and/or planes, where lines/planes were defined by specific anatomic landmarks or skull structures. An example of a constructed landmark is RT-abo, which is lying on a silhouette, halfway between the landmarks articulare (Ar) and gonion (Go). Based on their expertise, AUDAX classified landmarks into five classes with respect to their importance in cephalometric analyses. The 38 most important landmarks (class 5) are highlighted in Table 2. On the other hand, AUDAX also classified the landmarks into five classes with respect to the difficulty of their determination. The six most difficult to determine landmarks (class 5) are underlined in Table 2. All 72 denoted landmarks were used as the ground truth in our research. Figure 1b depicts a sample image from this private database, with 72 annotated cephalometric landmarks.

The K-fold validation technique was employed by utilizing data from this database to verify the detection methods. The K parameter was set to 3, thus dividing the private database randomly into 3 folds of the same size (i.e., each fold consists of 1565 unique images).

Table 2. A list of 72 cephalometric landmarks annotated in the AUDAX private database. All 19 landmarks from the ISBI public database are also annotated in this database (denoted encircled). The 6 most difficult to determine landmarks are underlined, while the 38 most important landmarks for the cephalometric analyses are bolded. A description of the landmarks and their significance can be found in [3].

−1a–Apex of lower incisor	**−1i—Lower incisal incisor**	**−6a—Apex of lower 1st molar**	**−6c—Cusp of lower 1st molar**
−6d—Distal side of lower 1st molar	**+1a—Apex of upper incisor**	**+1i—Upper incisal incisor**	**+6a—Apex of upper 1st molar**
+6c—Cusp of upper 1st molar	+6d—Distal side of upper 1st molar	**+St′—Upper Stomion**	**A—Point A**
A′—Point Soft A	**ANS—Anterior Nasal Spine**	**APocc—Anterior point of occlusion**	**Ar—Articulare**
B—Point B	**B′—Point Soft B**	**Ba—Basion**	Ci–Clinoidale
Co—Condylion	Col′—Columella	Cp—Condylion posterior	Cs—Condylion superior
D—Point D	DC—Point DC	ER—End Ramus	FMN—frontomaxillary nasal suture
Gl′—Glabella	**Gn—Gnathion**	Gn′—Point Soft Gnathion	**Go—Gonion**
Hy—Hyoid	Ir—Point Ir	L1—L1	**Li′—Lower lip**
LLi—Lower Lip inside	**Ls′—Upper lip**	**Me—Menton**	**Me′—Point Soft Menton**
N—Nasion	**N′—Soft Nasion**	NC—Nasal crown	**Or—Orbitale**
Pg—Pogonion	**Pg′—Point Soft Pogonion**	PM—Suprapogonion	**Pn′—Pronasale**
PNS—Posterior Nasal Spine	**Po—Porion**	**PPocc—Posterior point of occlusion**	**Pt—Pterygoid point**
R1–R1	R3–R3	Rh—Rhinion	RO—Orbital roof of orbital cavity
RT-abo—aboRamalTangent	**S—Sella Turcica**	Se—Entry of Sella	SE—Sphenoethmoidal point
Si—Floor of Sella	**Sn′—Subnasale**	SOr—Supraorbitale	Sp—Dorsum of Sella
−St′—Lower Stomion	Te—Temporale	**tGo—Constructed Gonion (tangent)**	Th′—Throat
U1—U1	ULi—Upper lip inside	W—Walker point	ZyO—Zy Orbit Ridge

3. Computational Methods

3.1. SpatialConfiguration-Net: A Summary

Our proposed landmark detection approach is based on the SpatialConfiguration-Net (SCN) neural network introduced in [10]. The SCN network is a fully convolutional NN and consists of two components, namely (i) local appearance and (ii) spatial configuration components. Both components generate a multidimensional heat map h:

$$h(\mathbf{x}) \in \mathbb{R}^{H \times W \times N}, \tag{1}$$

where $\mathbf{x}$ is a location vector within the heat map, H and W are the height and width of the heat map (also the size of the input image), and N denotes the number of heat map channels (also the number of targeted landmarks). A location of the n-th landmark is predicted as the location of the global maxima in the n-th heat map channel.

The local appearance component is a multi-scale pyramid style network that employs a series of convolutions and downsamplings to extract feature maps. These feature maps are then upsampled and integrated across different scales. An output of this component is the multidimensional or multichannel heat map h of dimension of $H \times W \times N$. Every channel of h can, therefore, be treated as a separate 2D heat map ($H \times W$) that estimates the location of a selected landmark (i.e., N channels for N landmarks).

The spatial configuration component downsamples, by a large factor, the heat map estimated by the local appearance component. It processes this heat map with another series of convolutions with larger kernels, and produces the new multichannel heat map, which is upsampled appropriately at the end. Afterwards, the heat maps from the spatial configuration component, h^{SC}, and from the local appearance component, h^{LA}, are merged into a new multidimensional heat map h by using the Hadamard product (i.e., element-wise product) as:

$$h(\mathbf{x}) = h^{LA}(\mathbf{x}) \circ h^{SC}(\mathbf{x}). \quad (2)$$

Figure 2 visualizes the above described procedure.

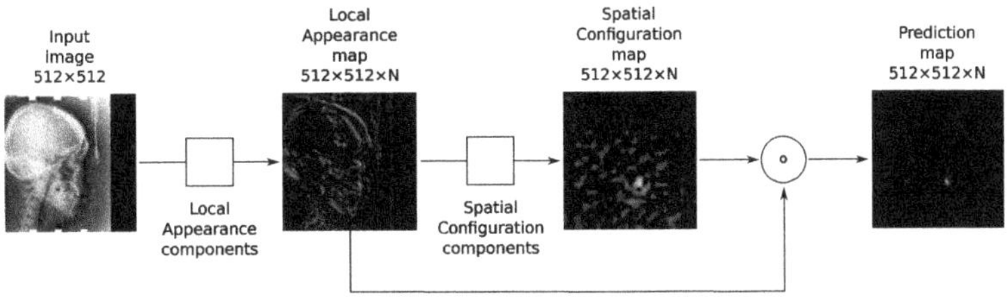

Figure 2. A rough block diagram of the SpatialConfiguration-Net. Depicted are ground plan views of maps (i.e., a single 2D map/channel is shown for a selected landmark).

The local appearance component was designed to learn an accurate landmark position based on local information. On the other hand, the spatial configuration component is aimed to discriminate between possible landmark locations using a larger or global context. An element-wise multiplication of both heat maps is an essential part of the SCN architecture. The latter enables the local appearance component to make multiple estimates for a landmark location across the image, while the spatial configuration component is allowed to selected between these estimates. The local appearance component can, thus, be focused on accurate position estimation without a global discrimination knowledge, while the spatial configuration component does not need to have an accurate landmark's position information, but it is focused on the global discrimination of the landmark's position.

3.2. Proposed SCN-EXT Method

The aim of this research is to develop an effective deep-learning-based method for detecting a large number of cephalometric landmarks from skull X-ray images. State-of-the-art cephalometric landmark detection methods such as [8–11] have proven very effective on a small number of landmarks. Our goal, however, is to upgrade the state-of-the-art appropriately, also for more challenging kinds of detection.

A substantial increase in the number of targeted landmarks requires, typically, an increase in a (convolutional) neural network's capacity. A trivial solution of increasing the number of filters for each convolution layer proved to have two drawbacks. First, doubling the number of filters squares the number of free parameters for most layers. Consequently, the memory requirements grow quadratically. Second, increasing the number of parameters typically makes the learning of an NN with the same training set and similar hyperparameters either unstable or prone to overfitting [12].

The considerable inflation of free parameters is particularly acute for the SCN network, as we have found through experimentation that this network learning has become very unstable. It should also be noted that an exhaustive fine-tuning of the initialization constants for particular SCN layers were carried out. It is expected that an additional fine-tuning of the SCN network would be required by larger expansion of the free parameters. However, the SCN network performed with high accuracy when detecting 19 cephalometric landmarks on testing images from the ISBI public database (see Section 2.2).

We wanted to take advantage of the high detection effectiveness of the SCN network, but, at the same time, we wanted to avoid re-evaluating (i.e., fine-tuning) the initialization constants if the SCN network capacity was increased significantly. Therefore, we propose the following SCN network extension, denoted as SCN-EXT, which increases the capacity of the NN by adding a series of new, but with the same hyperparameters, basic building blocks of the SCN network.

We constructed the SCN-EXT network by introducing J repetitions of the local appearance component into the SCN network, where each of these components was connected with an input image. Figure 3 depicts the basic elements and outputs of the SCN-EXT network. An output of the local appearance component is a multichannel heat map (dimensions of $H \times W \times N$), which is passed on to the input of the new spatial configuration component. We must, therefore, integrate J spatial configuration components into the SCN-EXT network, i.e., one for each local appearance component. The spatial configuration component also returns as an output of the matrix of dimension $H \times W \times N$ (i.e., spatial configuration map). Subsequently, combining the outputs of all J repetitions of a particular component follows. The J outputs of the local appearance components are summed simply into the final local appearance heat map. Similarly, the spatial configuration components' outputs are combined (see Figure 3). Finally, identical to the original SCN network, both the final local appearance and the final spatial configuration heat maps are merged, by using the Hadamard product, into a prediction map, which is then utilized for predicting landmarks' locations. The described procedure for constructing the prediction map h is written formally as:

$$h(\mathbf{x}) = \left(\sum_{j=0}^{J} h_j^{LA}(\mathbf{x})\right) \circ \left(\sum_{j=0}^{J} h_j^{SC}(\mathbf{x})\right), \quad (3)$$

where h_j^{LA} and h_j^{SC} denote heat maps of the j-th local appearance and the j-th spatial configuration component, respectively. It should be emphasized once again that, in the SCN-EXT network, we employed the basic components with the same hyperparameters from the SCN network (i.e., components were initialized with the recommended settings from [10]).

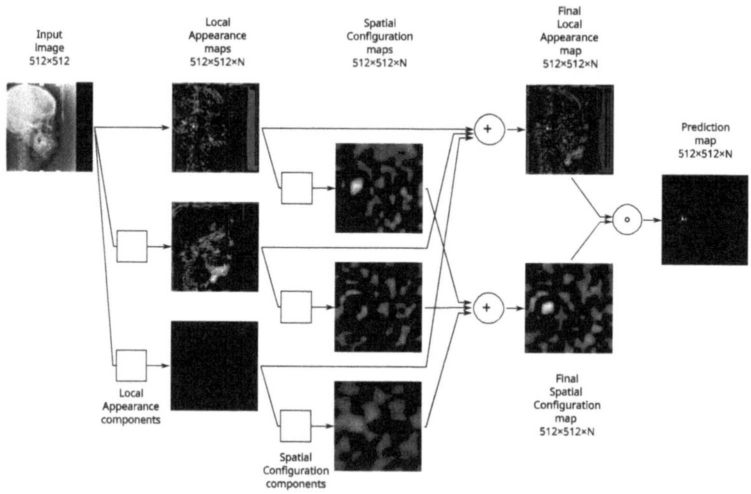

Figure 3. A rough block diagram of the proposed extended SpatialConfiguration-Net (SCN-EXT).

By adding $J - 1$ new local appearance and spatial configuration components, the proposed SCN-EXT network is able to learn $J^2 - 1$ functions more than the original SCN network. Each such component (i.e., neural network) has independent training parameters, and can, thus, learn a subset of the targeted landmarks. On the other hand, compared to the base SCN network, the number of free parameters in SCN-EXT grows linearly with the number of components used.

Landmarks in a training set are not separated into groups (e.g., with respect to an individual anatomical feature or with respect to the neighboring position), so a benefit of utilizing several components in the SCN-EXT is that they can optimize for self-determined and overlapping groups of landmarks. Each component (neural network) needs to estimate only a fraction of all the targeted landmarks, and multiple networks can cooperate on the same landmark.

An idea in our solution is similar to the so-called grouped convolutions, where the channels in a single convolution layer are grouped together. Each group of channels is processed by a separate set of convolution kernels without overlap between the groups. This has a similar advantage as our proposed approach: a moderate increase of free parameters despite a greater increase of convolutional filters. Increasing the capacity of the CNN network considerably, i.e., the ability to learn several new functions, by a small increase of the number of degrees of freedom (DoF), thus provides more stable learning by using the same training set.

4. Implementation Details and Evaluation Metrics

4.1. Implementation Details and CNN Training

First of all, we will describe image preprocessing and the preparation of training data, followed by an explanation about the training procedure.

Initially, each image was zero-padded along its shorter axis to make it square shaped. Afterwards, it was resampled to a size of 512 × 512 pixels. A variability in the training set was increased by an augmentation. The training images were augmented "on the fly" by random rotations (±5°), uniform scaling with a scaling factor selected randomly between 0.6 and 1.2, and intensity changes with a random factor from an interval [0.75, 1.3].

The ground truth heat maps were generated as instructed in [10]. Gaussian kernels were placed at known landmark positions. The Gaussian kernel values were multiplied by a constant $\gamma = 100$ to reduce training instabilities. The standard deviations of the kernels were the training parameters, where they were regularized by using L2-regularization with a weight of 20.

All our own implemented neural networks were trained by using the Adam optimization algorithm [13], with an initial learning rate of 1×10^{-4}. The learning rate was reduced by a factor of 0.5 every 50 epochs without loss improvement on the validation set. The training was limited to a maximum of 150 epochs.

Our software was implemented by using the Python programming language. The constructed and implemented deep neural networks were trained by using the TensorFlow software library. Originally, version 1.15 was employed, but later the code was ported to 2.x libraries (at the end, the models were trained in the 2.4 version library).

All experiments were conducted on a computer system with an AMD Ryzen Threadripper 2920X 12-Core processor, an NVidia Quadro GV100 graphical card with 32 GB of VRAM and 64 GB of physical RAM, and Samsung EVO 970 NVMe 1TB storage.

4.2. Evaluation Metrics

Evaluation metrics and the protocol prescribed for the ISBI public database [4,5] were employed to validate the cephalometric landmark detection methods in this study. The validation was based on the radial error (RE), calculated as the Euclidean distance $d()$ between the estimated, **EST**, and ground-truth landmark location, **GT** (i.e., the 2D point). A basic metric mean radial error (MRE) is derived from this error, where the MRE is calculated as the average of radial errors over L observed landmarks, which is written formally as

$$MRE = \frac{1}{L}\sum_{i=1}^{L} d(\mathbf{EST}_i, \mathbf{GT}_i), \tag{4}$$

where $\mathbf{EST}_i$ in $\mathbf{GT}_i$ denote the estimated and ground-truth locations for the i-th landmark. It should be stressed that L denotes the number of landmarks, and does not necessarily represent the number of different types of landmarks observed in each X-ray image (this is denoted as N in this article).

Two additional statistics of radial error were calculated besides the mean (and the standard deviation) in this research, namely, the median and the 90th percentile of radial error. All the mentioned measures can be estimated per landmark type, per image, or even per all landmark types and all images (i.e., over all landmarks in all images in the database). These metrics are presented either in pixels or in mm if the spatial resolutions of the images are known.

The next metric that has been introduced for the ISBI database is the successful detection rate (SDR), which evaluates the precision (i.e., the positive predictive value) of landmark detection with respect to the radial error. The metric SDR is assessed typically in respect to the radial error up to 2 mm (Class 1), 2.5 mm (Class 2), 3 mm (Class 3), and 4 mm (Class 4) from the ground-truth landmark position. It should be noted that we were unable to determine this metric for the AUDAX private database, because the spatial resolution information was not known for this database.

5. Results

First, we will describe the experiment by which we fine-tuned the SCN-EXT network architecture, and afterwards, we will present the results obtained by the detection of cephalometric landmarks on the ISBI and AUDAX databases.

5.1. SCN and SCN-EXT Architecture Determination

Our research is based on the SCN neural network. The implementation of this network is, to the best of our knowledge, not publicly available; therefore, based on the available information, we recreated the SCN network ourselves. We tested our own implemented SCN on the public ISBI database by using the (hyper)parameters reported in [10]. The SCN network had the following architecture. The local appearance component had 4 layers and 128 filters with 3×3 kernels. The spatial configuration component used a downsampling factor of 16 and included 128 filters of 11×11. In total, this network had around 7.90 million (M) trainable parameters. The SCN network with the described architecture was referred to in the sequel as "our implementation of the method".

Our proposed SCN-EXT solution is a generalization of the SCN architecture, with J-times repetition of local appearance and spatial configuration components (see Section 3.2). We determined the most acceptable SCN-EXT architecture by using the following simple experiment. This experiment was conducted on the AUDAX private database, whereas folds 2 and 3 formed the training set, while fold 1 was utilized as the testing set. According to the presented theory in Section 3, we integrated J repetitions of both components of the SCN network into the SCN-EXT network. If we had employed our fine-tuned SCN for this purpose, then the memory requirements would have become so high (even at small values of J) that this problem could not be solved with today's available hardware. Therefore, we utilized the following simplified SCN architecture for this experiment: (i) local appearance component: 4 layers and 32 filters with 5×5 kernels; and (ii) spatial configuration component: a downsampling factor equal to 16 and 32 filters with 11×11 kernels. Afterwards, the SCN-EXT networks were constructed by changing the number of repetitions of SCN network components, whereas parameter J was varied between 1 and 10 with step 1. It should be stressed that for $J = 1$, we are dealing with the original SCN network.

The results obtained by using different SCN-EXT architectures are summarized in Table 3. The number of repetitions (J) of the SCN architecture components is written next to the method name. For comparison, we also added in this table the results of the fine-tuned SCN architecture (see the first line). Three metrics are shown based on the radial error. All metrics were evaluated across all 72 cephalometric landmarks and across all 1565 testing images. The values are given in pixels, where a lower value indicates the more effective method. We added a number of trainable parameters in the last column. Marked in bold is the SCN-EXT architecture, i.e., SCN-EXT ($J = 6$), which was used in all subsequent experiments. We chose this network because it is a good compromise between effectiveness and training time. At the same time, this network is similar to the fine-tuned SCN with respect to the DoF (see the "trainable" column). As both networks have similar DoFs, in fact the SCN-EXT network ($J = 6$) has even 1 M lower DoF, all differences in the results can be attributed to changes in the CNN architecture, and not to a raw increase of the DoF (as in the case if we would utilize SCN-EXT with $J = 9$).

Table 3. Effectiveness of different SCN-EXT architectures on cephalometric landmark detection on fold 1 of the AUDAX private database. The column MRE denotes the mean and standard deviation of the radial error, while columns $PCTL_{50}$ and $PCTL_{90}$ denote the 50th (i.e., median value) and 90th percentile of the radial error, respectively. All values are in pixels ("px"). The column "trainable" presents the number of trainable parameters in millions.

Method	MRE (px)	$PCTL_{50}$ (px)	$PCTL_{90}$ (px)	Trainable
SCN †	11.56	6.70	24.91	7.90 M
SCN-EXT, $J = 1$	12.35	7.21	26.44	1.15 M
..., $J = 2$	11.80	6.90	25.25	2.29 M
..., $J = 3$	11.57	6.75	24.72	3.44 M
..., $J = 4$	11.56	6.73	24.68	4.58 M
..., $J = 5$	11.54	6.69	24.57	5.73 M
..., $J = 6$	**11.36**	**6.66**	**24.31**	**6.88 M**
..., $J = 7$	11.42	6.67	24.30	8.02 M
..., $J = 8$	11.35	6.54	24.20	9.17 M
..., $J = 9$	11.26	6.57	24.05	10.31 M
..., $J = 10$	11.48	6.60	24.48	11.46 M

†—Our implementation of the method.

5.2. ISBI Public Database

Initially, the effectiveness of our proposed SCN-EXT method, designed primarily for cephalometric landmark detection, was assessed on the ISBI public database. We used the prescribed methodology and established metrics [5]. The mean and standard deviation of the radial error were calculated over all 19 cephalometric landmarks and over all testing images. In addition, the successful detection rate (SDR) metric was evaluated for the four prescribed classes. The results for testing set 1 are gathered in Table 4, while Table 5 summarizes the obtained results for testing set 2.

Table 4. Effectiveness of cephalometric landmark detection methods on the public ISBI database: testing set 1. The column MRE denotes the mean and standard deviation of the radial error, while the SDR columns denote the successful detection rate (in %) for the four specified classes.

Method	MRE (mm)	SDR (%) 2 mm	SDR (%) 2.5 mm	SDR (%) 3 mm	SDR (%) 4 mm
Li et al. [9]	1.04 ± N/A	88.49	93.12	95.72	98.42
SCN [10] †	1.08 ± 1.08	87.30	91.40	94.25	97.33
Song et al. [11]	1.08 ± N/A	86.40	91.70	94.80	97.80
SCN-EXT	**1.13 ± 1.11**	**85.61**	**90.60**	**93.96**	**97.44**
Chen et al. [8]	1.17 ± N/A	86.67	92.67	95.54	98.53
Chen et al. [8] †	1.30 ± 2.07	83.65	90.70	94.81	97.86
Lindner et al. [5]	1.67 ± 1.48	73.68	80.21	85.19	91.47
SCN [10] ‡	N/A	73.33	78.76	83.24	89.75
Ibragimov et al. [5]	N/A	71.72	77.4	81.93	88.04

N/A—Data not available. †—Our implementation of the method. ‡—Results reported just for the merged testing set 1 and 2.

The effectiveness of the state-of-the-art methods were added to the tables as well. Implementations of these methods were not publicly available; therefore, we just summarized the results published by the authors of the methods. We reimplemented only two state-of-the-art methods successfully. The remaining methods were either basically too ineffective (e.g., methods [6,7]), or it was very difficult to scale them to the problem of 72 cephalometric landmarks' detection (e.g., method [11]), or method descriptions were not comprehensive enough to be able to reproduce them accurately (e.g., method [9]). Additionally, the results of our methods' implementations are presented in the tables, where they are marked by † next to the method name.

The methods in both tables are arranged according to the decreasing value of the MRE metric. Let us emphasize that a lower MRE value indicates a higher detection effectiveness of the method, which means that the better methods are at the top of the tables.

Table 5. Effectiveness of the cephalometric landmark detection methods on the public ISBI database: testing set 2. See Table 4 for denotations.

Method	MRE (mm)	SDR (%) 2 mm	SDR (%) 2.5 mm	SDR (%) 3 mm	SDR (%) 4 mm
SCN [10] †	1.41 ± 1.40	74.84	81.42	86.89	94.47
Li et al. [9]	1.43 ± N/A	76.57	83.68	88.21	94.31
SCN-EXT	**1.47 ± 1.44**	**74.53**	**82.21**	**87.21**	**93.68**
Chen et al. [8]	1.48 ± N/A	75.05	82.84	88.53	95.05
Chen et al. [8] †	1.65 ± 2.22	71.79	80.32	86.21	93.84
Song et al. [11]	1.54 ± N/A	74.00	81.30	87.50	94.30
Lindner et al. [5]	1.92 ± 1.24	66.11	72.00	77.63	87.42
SCN [10] ‡	N/A	73.33	78.76	83.24	89.75
Ibragimov et al. [5]	N/A	62.74	70.47	76.53	85.11

N/A—Data not available. †—Our implementation of the method. ‡—Results reported just for the merged testing set 1 and 2.

5.3. *AUDAX Private Database*

The effectiveness of our proposed SCN-EXT method was also assessed on the AUDAX private database. On this database, we applied the threefold validation technique, where there were 1565 images in each fold and 72 cephalometric landmarks in each image. The results obtained on the individual folds were merged, and, afterwards, summarized with various statistics calculated over all images and over all cephalometric landmarks. We calculated the mean radial error and the 50th and 90th percentiles of the radial error. The spatial resolution for the AUDAX database is not known; therefore, all results are given in pixels. The calculated metrics are gathered in Table 6. In addition to our proposed SCN-EXT detection method, this table also presents the results of our implementations of two state-of-the-art methods. The methods in the table are arranged according to the decreasing value of the MRE metric. Based on publicly available information, we also reimplemented the method by Li et al. [9], but, with the calculated MRE of about 34 pixels and the median radial error around 25 pixels, we found that our attempt was completely unsuccessful.

The effectiveness of the methods in Table 6 was also assessed with the nonparametric Friedman's statistical test [14] at a 0.05 significance level. The calculated p-value was equal to 0, which indicates that not all the methods' medians are equal. The proposed SCN-EXT method had the lowest mean rank of 1.88, followed by the SCN method with the mean rank of 1.94, and the method of Chen et al. [8] had the highest mean rank of 2.18. Let us evoke that the lower mean rank correlates with the lower radial error, and, consequently, with the higher effectiveness of the method. Subsequently, we conducted a multiple comparison test of mean ranks, i.e., a pairwise comparison of methods. This analysis pointed out that all three compared methods have significantly different mean ranks. On this basis, we argue that our proposed approach has proven overall to be the most effective detection method on the challenging AUDAX database.

Table 6. Effectiveness of the better cephalometric landmark detection methods on the private AUDAX database. The column MRE denotes the mean and standard deviation of the radial error, while columns $PCTL_{50}$ and $PCTL_{90}$ denote the 50th (i.e., median value) and 90th percentiles of the radial error, respectively. All values are in pixels.

Method	MRE (px)	$PCTL_{50}$ (px)	$PCTL_{90}$ (px)
SCN-EXT	**11.26 ± 17.51**	**6.52**	**24.13**
SCN [10] †	11.57 ± 18.71	6.70	25.10
Chen et al. [8] †	12.19 ± 15.02	8.36	25.00

†—Our implementation of the method.

In the sequel, we extracted the metrics from the obtained results only for those 19 landmarks that are also annotated in the public ISBI database. The mean and standard deviation of the radial error was calculated for each landmark and each compared method separately over all images (i.e., 4695 images). These metrics are accumulated in Table 7. The effectiveness of the methods was then assessed by Friedman's statistical test (0.05 significance level), and by a multiple comparison test of mean ranks. In the table next to the MRE value, we wrote in parentheses the order of methods with respect to the mean rank (value 1 indicates the most effective and value 3 the least effective method), where we denoted by an asterisk whether the differences in results are statistically significant. Our proposed method proved to be the most accurate by 15 landmarks and the second best by 4 landmarks, which is notably better than the compared methods. Improvements were statistically significant for six landmarks. Finally, we calculated the MRE over all 19 landmarks (see the row "all landmarks" in the table). The effectiveness of our proposed detection method was statistically significantly higher by at least 3% than for the compared methods. The SCN method was shown to be the second most effective, followed by the method by Chen et al. [8].

Table 7. Effectiveness of the compared methods on the AUDAX database. Considered are only landmarks from the ISBI database. The mean and standard deviation of the radial error are presented in pixels. A number and * in () denote the method's rank and statistically significant difference. **Better results are marked in bold.**

Landmark	SCN-EXT (px)	SCN [10] † (px)	Chen et al. [8] † (px)
−1i–Lower incisal incisior	$\mathbf{5.06 \pm 7.34}^{(1)}$	$5.09 \pm 7.48^{(2)}$	$7.45 \pm 7.72^{(3)}$
+1i–Upper incisal incisior	$\mathbf{4.55 \pm 6.72}^{(1)}$	$4.55 \pm 6.83^{(2)}$	$8.33 \pm 7.77^{(3)}$
ANS–Anterior Nasal Spine	$\mathbf{8.96 \pm 10.88}^{(1,*)}$	$9.35 \pm 11.55^{(2)}$	$12.16 \pm 19.39^{(3)}$
Ar–Articulare	$\mathbf{8.07 \pm 9.38}^{(1,*)}$	$8.44 \pm 9.69^{(2)}$	$9.56 \pm 8.14^{(3)}$
Gn–Gnathion	$\mathbf{7.11 \pm 6.08}^{(1,*)}$	$7.29 \pm 6.22^{(2)}$	$8.04 \pm 6.17^{(3)}$
Go–Gonion	$9.40 \pm 8.50^{(2)}$	$11.05 \pm 10.15^{(3)}$	$\mathbf{8.81 \pm 7.11}^{(1)}$
Li'–Lower lip	$\mathbf{4.51 \pm 6.77}^{(1)}$	$4.66 \pm 12.77^{(2)}$	$7.01 \pm 6.80^{(3)}$
Ls'–Upper lip	$\mathbf{4.49 \pm 6.30}^{(1)}$	$4.63 \pm 8.68^{(2)}$	$7.16 \pm 6.50^{(3)}$
Me–Menton	$\mathbf{6.77 \pm 6.52}^{(1,*)}$	$6.94 \pm 6.63^{(2)}$	$7.86 \pm 6.17^{(3)}$
N–Nasion	$7.31 \pm 10.21^{(2)}$	$\mathbf{7.29 \pm 10.24}^{(1)}$	$9.12 \pm 10.09^{(3)}$
Or–Orbitale	$12.22 \pm 14.29^{(2)}$	$\mathbf{12.15 \pm 14.20}^{(1)}$	$12.62 \pm 11.92^{(3)}$
Pg–Pogonion	$\mathbf{7.17 \pm 9.34}^{(1)}$	$7.24 \pm 9.49^{(2)}$	$9.62 \pm 8.98^{(3)}$
Pg'–Point Soft Pogonion	$\mathbf{9.02 \pm 15.80}^{(1)}$	$9.30 \pm 31.05^{(2)}$	$9.88 \pm 10.73^{(3)}$
PNS–Posterior Nasal Spine	$\mathbf{9.14 \pm 7.64}^{(1)}$	$9.31 \pm 7.87^{(2)}$	$11.65 \pm 8.61^{(3)}$
Po–Porion	$\mathbf{13.44 \pm 15.34}^{(1)}$	$13.89 \pm 16.21^{(2)}$	$14.89 \pm 12.43^{(3)}$
S–Sella Turcica	$\mathbf{4.85 \pm 3.58}^{(1,*)}$	$4.99 \pm 3.76^{(2)}$	$5.20 \pm 3.60^{(3)}$
Sn'–Subnasale	$\mathbf{5.68 \pm 5.35}^{(1,*)}$	$5.84 \pm 6.77^{(2)}$	$8.02 \pm 6.12^{(3)}$
SS–Subspinale (Point A)	$\mathbf{11.02 \pm 12.70}^{(1)}$	$11.28 \pm 13.41^{(2)}$	$14.16 \pm 12.37^{(3)}$
SM–Supramentale (Point B)	$12.88 \pm 17.02^{(2)}$	$\mathbf{12.92 \pm 16.90}^{(1)}$	$14.08 \pm 15.03^{(3)}$
All landmarks	$\mathbf{7.98 \pm 10.57}^{(1,*)}$	$8.22 \pm 12.82^{(2)}$	$9.77 \pm 10.29^{(3)}$

†—Our implementation of the method.

We gathered in Tables 8 and 9 the ten most accurately and the ten least accurately detected cephalometric landmarks by using our proposed SCN-EXT method. Among the ten best detected landmarks, there are as many as six landmarks (in Table 8 they are circled), which are also annotated in the ISBI database. The Th′ point from the soft tissue of the throat was detected with the largest MRE error, which differs significantly from others (see Table 9). The reason is that the throat is not fully visible on all cephalograms and, therefore, an expert annotated the Th′ point very inconsistently (i.e., Th′ was annotated only approximately). If the Th′ point was excluded from the metric calculation, the MRE for the SCN-EXT method decreased by about 0.7 pixels to 10.57 ± 13.93 pixels (see also Table 6). Similarly, the median radial error decreased to 6.44 pixels (previously 6.52) and the 90th percentile of the radial error to 23.21 pixels (previously 24.13).

Table 8. Ten cephalometric landmarks from the AUDAX database detected most accurately by the SCN-EXT method. For denotations, see Tables 2 and 6.

Landmark	MRE (px)	$PCTL_{50}$ (px)	$PCTL_{90}$ (px)
Ls′–Upper lip	4.49 ± 6.30	3.14	8.49
Li′–Lower lip	4.51 ± 6.77	3.30	8.53
+1i–Upper incisal incisor	4.55 ± 6.72	3.00	8.30
Pn′–Pronasale	4.61 ± 7.13	3.46	9.20
S–Sella Turcica	4.85 ± 3.58	3.97	9.67
APocc–Anterior point of occlusion	4.95 ± 5.27	3.66	9.60
Si–Floor of Sella	4.97 ± 4.61	3.90	9.97
−1i–Lower incisal incisor	5.06 ± 7.34	3.55	9.90
B′–Point Soft B	5.22 ± 6.69	3.64	10.06
Sn′–Subnasale	5.68 ± 5.35	4.41	11.58

Table 9. Ten cephalometric landmarks from the AUDAX database detected least accurately by the SCN-EXT method. For denotations, see Table 6.

Landmark	MRE (px)	$PCTL_{50}$ (px)	$PCTL_{90}$ (px)
SOr–Supraorbitale	15.81 ± 20.29	7.99	43.83
ZyO–Zy Orbit Ridge	16.90 ± 15.79	11.84	38.37
Te–Temporale	17.03 ± 18.05	12.39	35.41
Ir–Point Ir	17.09 ± 17.22	12.15	37.25
R1–R1	17.25 ± 14.82	13.25	35.61
Gn′–Point Soft Gnathion	18.59 ± 22.10	10.95	46.41
Gl′–Glabella	19.21 ± 20.64	11.91	46.66
R3–R3	19.58 ± 16.82	14.69	41.66
Rh–Rhinion	24.08 ± 33.69	10.49	81.31
Th′–Throat	60.15 ± 76.76	28.97	177.48

6. Discussion

In this study, we upgraded the state-of-the-art SCN neural network to the SCN-EXT network by adding the J repetitions of both the local appearance (LA) component and the spatial configuration (SC) component into the original SCN architecture. All J replicates of each component were summed up simply, and both sums were, finally, combined by using the Hadamard product. By modifying the architecture in this way, we increased the capacity, as the new SCN-EXT network is able to learn $J^2 - 1$ more transformation functions than the basic SCN network. It is completely trivial that if we add J copies of LA and SC components, then the capacity of such a modified network will, of course, increase compared to the capacity of the original SCN network (if the same LA and SC components are utilized). However, the contribution of our approach is that by J-times repeating and merging the simpler LA and SC components, we can maintain approximately the same DoF of the new SCN-EXT network as has the original SCN network with the more complex LA and SC components, while we simultaneously increase the capacity and learning ability of the SCN-EXT, respectively. The latter is especially acute if processing and memory resources are limited; namely, training the large models (i.e., with large DoF) requires powerful computing units, a large learning set, and a large primary memory.

This research was focused on the problem of detecting many cephalometric landmarks on diverse lateral skull X-ray images. The SCN-EXT network was designed primarily for this purpose. We have shown experimentally (see the Section 5) that the SCN-EXT network components learn well to predict landmark locations. In our current solution, we do not supervise a training by forcing individual components to learn how to localize a specific subset of landmarks. The latter would be achieved, for example, by adding the $L1$ regularization term for sparsity into the training, which could be one of the future research guidelines.

The final architecture of the SCN-EXT network was determined according to the capacity and DoF of the original SCN network. The SCN network was fine-tuned to detect 19 cephalometric landmarks in the ISBI public database. The LA and SC components utilized there were used as the basis in our work. The goal on the private AUDAX database was to localize 72 cephalometric landmarks; therefore, we modified the architecture of the SCN network only slightly, namely, such that the LA and SC components were able to process inputs with 72 channels. The SCN network that aimed for a detection of 19 cephalometric landmarks (ISBI database) had 6.20 M trainable parameters, while the DoF increased to 7.90 M in the case of detecting 72 landmarks (AUDAX database). The SCN-EXT architecture was determined by a simple experiment on the AUDAX database (see Section 5.1). We varied the number of replicates, J, of the LA and SC components, and monitored the MRE by cephalometric landmarks' detection. Much simpler LA and SC components were applied than in the original SCN. Finally, we chose the SCN-EXT architecture with $J = 6$ repetitions of both components with respect to the hypotheses set out in this study. The SCN-EXT network had 6.88 M trainable parameters when detecting 72 landmarks (AUDAX database), while the DoF decreased to 4.16 M if this architecture was adapted for the ISBI database (i.e., reducing the number of channels). It can be noticed easily that the SCN-EXT network had, on both databases, much fewer trainable parameters than the original SCN.

In order to compare the results of our proposed SCN-EXT method with the results of related works, we reimplemented the SCN method and the method by Chen et al. [8] successfully. We also implemented the method by Li et al. [9], but the results, obtained with our implementation of this method, differed greatly from those reported (see the previous section). We deduced that a reason for the failure to reproduce the method is as follows: the method by Li et al. [9] models each landmark as a graph node. Each node is associated with the landmarks' positions and a feature vector that is extracted from a processed image at that position. The feature vector processing is conducted by using the HRNet18 backbone convolutional network. This method consists of two stages. The first stage estimates a global perspective transformation to align the mean positions of landmarks, constructed from the training data with the specific image. Afterwards, the second stage refines local landmark locations. The estimated global perspective transformation did not improve the landmarks' locations regularly, but, rather, it distorted them. A network that predicted nine free parameters of the perspective transformation matrix was described in [9] explicitly. However, DeTone et al. argued in [15] that such approach is unreliable and difficult to train perspective transformations. Therefore, they suggested applying the four-point estimation approach instead. It is unclear, though, how this four-point estimation would be applied for the landmark detection. The reason for the ineffectiveness of this method was, consequently, sought in the poorly estimated perspective transformations. As mentioned in the Introduction, the method by Song et al. [11] does not scale well to a larger number of cephalometric landmarks and training images. The authors validated their approach on the ISBI public database (i.e., on 19 landmarks and 150 testing images). They reported that a registration of a single testing image to training images was completed in approximately 20 min. In the AUDAX database, there were 3130 training images per one fold. We estimated that registration in this case would require about 20 times more processing time, i.e., about 400 min per one testing image. In total, this would mean 3 folds $\times$ 1565 images $\times$ 400 min per image = 1,878,000 min, or around 1304 days, to carry out the registration. The latter, of course, is not acceptable, so we have not implemented this method. The remaining methods from Table 4 were around 40% behind the SCN method in terms of effectiveness, and were, therefore, not included in the comparison on the private AUDAX database.

First, let us analyze the results on the ISBI public database. The effectiveness of the proposed SCN-EXT method is comparable to the effectiveness of state-of-the-art cephalometric landmark detection methods. The SCN-EXT is, on testing set 1, less effective by about 8.65% than the best method by the authors Li et al. [9], and on testing set 2 by about 4.26% than the best SCN method (see Tables 4 and 5). We were unable to reproduce the results of [9], because important implementation details are missing in this method's presentation. Undoubtedly, one of the reasons for the lower effectiveness of our SCN-EXT method is that the architecture was established by using the AUDAX database (and not the ISBI data on which the method was actually applied). It should be noted that the DoF of the SCN-EXT method was almost one-third smaller than the DoF of the SCN method. It can also be seen on testing set 2 that the SCN and SCN-EXT methods have very similar SDR metrics. A great similarity between the methods was also perceived on testing set 1. A reason for the higher MRE of the SCN-EXT method is, therefore, attributed to those landmarks for which the SDR was >4 mm (i.e., incorrectly detected landmarks were detected more erroneously than in the SCN method). Finally, let us emphasize that the ISBI database is a small database with a small learning set (150 images), and with only 250 testing images divided into two sets.

Let us continue with an analysis of the results on the AUDAX private database. This database is very challenging, as it contains 4695 (testing) images, divided into 3 folds, in 287 very different sizes. A goal was to localize 72 cephalometric landmarks in each image. Spatial image resolution data were not available. To the best of our knowledge, this is the first such public or private database with a large number of X-ray images and a larger number of landmarks on which the cephalometric landmark detection methods have been verified. Taking into account all 72 cephalometric landmarks, our proposed SCN-EXT method proved to be superior compared to other state-of-the-art methods. It was more effective than the second-ranked SCN method by about 2.68% (see Table 6). The differences and rankings were confirmed statistically significantly by the nonparametric Friedman's test, and by the multiple comparison test of mean ranks. If we took into consideration from the set of all cephalometric landmarks only those 19 landmarks that were also annotated in the ISBI public database, then the SCN-EXT method this time again proved to be statistically significantly the best method. It surpassed the second-best SCN method by about 2.92% (see Table 7). A similar conclusion was drawn if we compared methods at the level of an individual cephalometric landmark. In this case, the SCN-EXT method was demonstrated to be the more effective method on 15 out of the 19 landmarks, and the second best on 4 landmarks. Afterwards, we arranged the detection effectiveness for the mentioned 19 landmarks with respect to the detection effectiveness for all 72 landmarks on the AUDAX database, where only our SCN-EXT method was observed. It was discovered that as many as 6 landmarks ranked among the top ten (even in the top three, see Table 8), 10 landmarks among the top twenty, and 15 landmarks among the top thirty-five most accurately detected cephalometric landmarks. The less accurately localized were the landmarks point A, orbitale, point B, and porion, as the least accurately detected landmark in 52nd place. On this basis, we argue that the ISBI database consists of 19 relatively easier to detect cephalometric landmarks. On the other hand, the AUDAX database can be said to contain at least 33 cephalometric landmarks, which are more difficult to localize than landmarks in the ISBI database. The latter makes the AUDAX database much more demanding than the ISBI database.

Figure 4 depicts the qualitative result of cephalometric landmarks' detection by using our proposed SCN-EXT method on the AUDAX private database. Seventy-two estimated (denoted by a red x) and ground-truth (blue circle) cephalometric landmarks are superimposed on the skull X-ray image. The predicted and correct location of the landmarks are connected by the green line, where the following applies: the shorter the line, the lower the radial error. It can be noticed that, with the exception of the point on the throat, all the remaining cephalometric landmarks were localized extremely accurately.

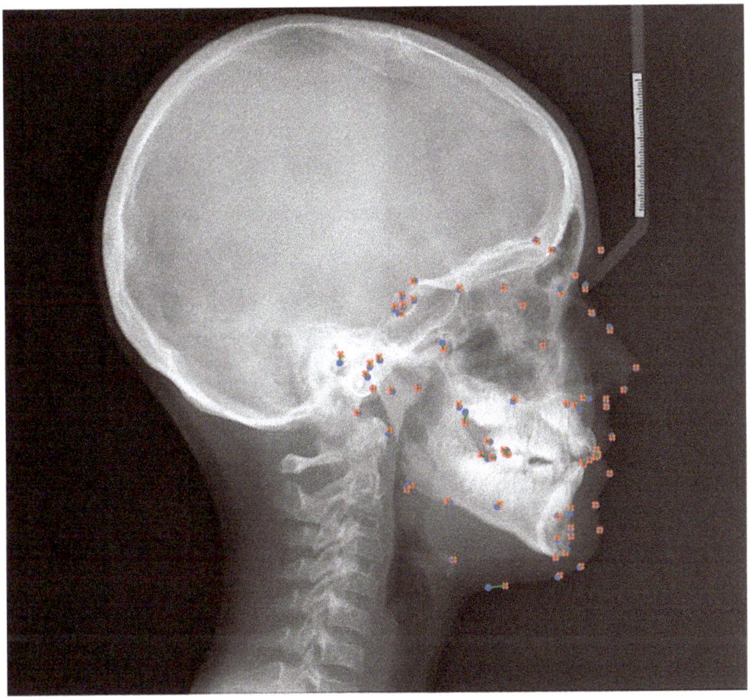

Figure 4. Sample detection result, superimposed on the X-ray image from the AUDAX private database. Cephalometric landmarks were determined by the proposed SCN-EXT method. Estimated landmarks are denoted by a red x, while ground-truth locations are superimposed as blue circles.

The rater's annotations were also analyzed on the AUDAX database. We wanted to find out the positions of which landmarks varied the most on the skull, and, whether the results obtained with our SCN-EXT method were consistent with these findings; accordingly, if the position of the landmark varied slightly on the skull and whether this made our method more accurate, and vice versa. Just a few findings are presented in the sequel, as this analysis is not the main goal of our research. We thus conducted a statistical analysis of skull shapes on the AUDAX database. Seventy-two annotated cephalometric landmarks from all 4695 images were utilized as an input. The aim of this analysis was to determine how the locations of cephalometric landmarks differ (vary, deviate) in the population (i.e., among patients), and how this influenced landmark detection effectiveness. We carried out a so-called generalized Procrustes analysis [16,17]. In each image, the locations of cephalometric landmarks were compensated by translation, scaling, and rotation (i.e., by a rigid transformation), resulting in a mean skull shape (and corresponding mean landmarks' locations) in the Procrustes space. Subsequently, we fitted the Procrustes mean model to the annotated cephalometric landmarks in each image by using an approach from [18], followed by the calculation of the radial error between the fitted model landmarks and the ground-truth landmarks. This error was summarized for each cephalometric landmark over all images with various statistics (i.e., mean, standard deviation, median, and the 75th percentile). It was discovered that the following 10 cephalometric landmarks have the lowest variability, namely, the landmarks PNS, APocc, W, S, Se, Ci, LLi, +St′, −St′, and PPocc (see Table 2 for denotations). The ten landmarks with the higher deviation from the Procrustes mean model are the landmarks Go, B, N′, Gn′, tGo, Ba, Gl′, Rh, Hy, and Th′, which is the overall highest variability landmark. Both lists remained the same regardless of any statistics (e.g., mean, median, etc.) used in the comparison.

Finally, we evaluated the influence of variability on the cephalometric landmark detection. We calculated the correlation between the landmark variability and detection effectiveness by using the SCN-EXT method. For both quantities, we used data regarding the points order, once in respect to the variability, the second in respect to the detection effectiveness. There was a positive correlation between the two quantities (the correlation coefficient equaled 0.505 with a p value 5.99×10^{-6}). To sum up, the less the landmark varied, the more accurately it was detected, and vice versa. These findings are also consistent with the importance of landmarks for cephalometric analyses as defined by the AUDAX company (see Table 2). With the exception of the Gl′ landmark, all the remaining nine poorly localized landmarks (see Table 9) are less important for the cephalometric analyses. Similarly, all 10 accurately localized landmarks (see Table 8) are more important for the cephalometric analyses.

The landmark on the throat soft tissue, Th′, with the MRE error of more than 60 pixels, was detected the least accurately. This MRE is almost 2.5 times higher than for the second-least accurately detected landmark, Rh. For the cephalometric analyses conducted by the AUDAX company, the landmark Th′ defines just a point where a face profile ends at the bottom. The landmark Th′ has no other meaning in these analyses, and, consequently, it was annotated very carelessly. Figure 5 depicts three examples of Th′ landmark annotation and localization by the SCN-EXT method. It can be noticed that Th′ was annotated on three completely different parts of the throat (see blue circles). Accordingly, this means a poorer ability to learn this landmark and a higher radial error (see the green lines). To illustrate, if we omitted the Th′ landmark from the statistics, then the MRE for the SCN-EXT method decreases from 11.26 pixels (see Table 6) to 10.57 pixels, or decreases by 6.13%.

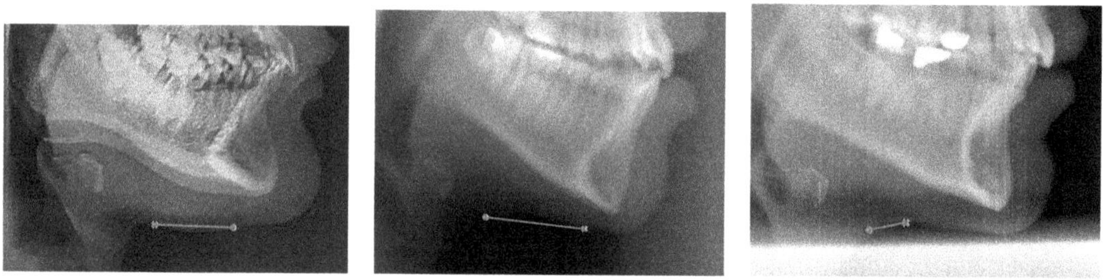

Figure 5. The worst-detected landmark Th′ by using the SCN-EXT method: three examples from the AUDAX database. Estimated landmarks are denoted by a red x, while ground-truth locations are superimposed as blue circles.

The CNN training was computationally demanding. The hardware utilized in this study was presented in Section 4.1. On the ISBI database, the training to detect 19 cephalometric landmarks took about 72 min for 150 epochs, or about 29 s per epoch (on GPU). The trained network conducted an inference in around 0.76 s per image on the CPU or in around 0.08 s per image on the GPU. On the AUDAX database, however, the training on GPU took about 2480 min for 150 epochs, or about 992 s per epoch. The trained network localized 72 cephalometric landmarks in around 1.02 s per image on the CPU or in around 0.14 s per image on the GPU.

7. Conclusions

By developing a new method for localizing cephalometric landmarks, we solved a concrete problem from industry in this research. The existing methods have been adapted and tested to detect only 19 landmarks; however, in our work we have addressed the problem of detecting 72 cephalometric landmarks based on industry needs. A large number of accurately detected landmarks on skull X-ray images is a prerequisite for any quality cephalometric analysis. In this study, we upgraded the SpatialConfiguration-Net neural network (SCN), which is one of the state-of-the-art methods for localizing cephalometric landmarks in X-ray images. The SCN architecture was modified by the integration of several repetitions of simpler local appearance and spatial configuration components, with which we increased the capacity of such a modified network (i.e., the SCN-EXT network) with virtually unchanged degrees of freedom (DoF) compared to the original SCN network with the more complex components. Primarily, the SCN-EXT network was designed for localizing a large number of cephalometric landmarks in diverse skull X-ray images.

On the small ISBI public database with 250 testing images, captured by the same X-ray device and with 19 cephalometric landmarks, our, albeit non-tuned SCN-EXT method, was, in terms of effectiveness, just slightly behind the state-of-the-art methods. On the other hand, our fine-tuned SCN-EXT method was statistically significantly the most accurate method on the much more demanding AUDAX database with 4695 highly variable testing images (various X-ray devices!) and with 72 cephalometric landmarks. The improvement of the proposed method was statistically significant, even if we considered out of all 72 cephalometric landmarks only those 19 landmarks that are also in the ISBI database. We also confirmed that the detection accuracy was correlated positively with the importance of landmarks for cephalometric analyses.

An aim of this research was indeed to develop a state-of-the-art cephalometric landmark detection method, but not at the expense of a raw increase of neural network capacity by increasing DoF (e.g., by the addition of more filters, etc.). The presented results in this study were, namely, obtained by using the SCN-EXT network, which had 13% (on the AUDAX database) or 33% (on the ISBI database) fewer free parameters than the original SCN network. Maintaining DoF while increasing network capacity is important, especially for a small learning set and limited computer resources.

Possible improvements to our approach are seen in the use of a more sophisticated augmentation of learning set and in the use of transfer learning. We expect, reasonably, also an improvement in the case if we integrate $J = 9$ or more repetitions of the local appearance and spatial configuration components to the SCN-EXT network, which would indeed increase DoF greatly. For the sake of a fair comparison with the state-of-the-art methods, we have not conducted any of the abovementioned in this study, so these may provide guidelines for future research.

In addition to lateral skull X-ray images, we also have an option of capturing frontal skull X-ray images. This is complementary information that allows complementary cephalometric analyses. One of the future research directions will, therefore, be focused on adapting our method for also localizing cephalometric landmarks on the frontal skull X-ray images.

Finally, let us mention that our detection algorithm is already employed in a clinical practice as a part of a bigger software product. Accurately determined landmarks on the skull X-ray images represent the input for every cephalometric analysis. Automatic localization of 72 cephalometric landmarks undoubtedly disburdens the orthodontist greatly, as manual detection of landmarks means routine and time-consuming work. Nevertheless, he should be aware that, similar to other software tools in clinical practice, our algorithm also does not work 100% accurately. Our trained model is well suited to support and aid manual cephalometric landmarks' annotation, but is not suited for fully automated systems. Manual validation is recommended, and manual correction may be required, based on final application requirements. For this reason, the orthodontist should be able to inspect, and possibly correct, the locations of automatically detected landmarks. Such functionality is, of course, built into the abovementioned software product. The user experiences of orthodontists with our algorithm are very positive. We conclude with one of the orthodontist's responses: "I conducted the first analysis. I have not used automated tracing for 3 years, but I saw that it is very improved. Landmarks are set at 99% ideally. Very good".

Author Contributions: Conceptualization, B.P. and M.Š.; methodology, B.P. and M.Š.; software, M.Š., G.S. and B.P.; validation, B.P., M.Š. and G.S.; formal analysis, B.P.; investigation, B.P.; resources, B.P.; data curation, B.P.; writing—original draft preparation, B.P., M.Š. and G.S.; writing—review and editing, B.P., M.Š. and G.S.; visualization, G.S. and M.Š.; supervision, B.P.; project administration, B.P.; funding acquisition, B.P. All authors have read and agreed to the published version of the manuscript.

Funding: This study was supported by the Slovenian Research Agency (Contract P2-0041).

Institutional Review Board Statement: Not applicable.

Informed Consent Statement: Not applicable.

Data Availability Statement: The source code used in this paper is publicly available from GitHub https://github.com/MartinSavc/SpatialConfNetExt (accessed on 13 April 2022). The ISBI database [4,5] is available from http://www-o.ntust.edu.tw/~cweiwang/ISBI2015/challenge1/ (accessed on 13 April 2022). The AUDAX database is not publicly available due to data privacy concerns regarding medical data. Data might be available on request from the AUDAX company (https://www.audaxceph.com (accessed on 13 April 2022)).

Acknowledgments: We gratefully acknowledge the valuable scientific and professional contribution of Peter Kobal, from the AUDAX company, Slovenia, who provided us with skull X-ray images and cephalometric landmark annotations (i.e., with the AUDAX database).

Conflicts of Interest: The authors declare no conflict of interest. The funders had no role in the design of the study; in the collection, analyses, or interpretation of data; in the writing of the manuscript, or in the decision to publish the results.

References

1. Douglas, T. Image processing for craniofacial landmark identification and measurement: A review of photogrammetry and cephalometry. *Comp. Med. Imag. Graph.* **2004**, *28*, 401–409. [CrossRef] [PubMed]
2. Durao, A.; Pittayapat, P.; Rockenbach, M.; Olszewski, R.; Ng, S.; Ferreira, A.; Jacobs, R. Validity of 2D lateral cephalometry in orthodontics: A systematic review. *Prog. Orthod.* **2013**, *14*, 31. [CrossRef] [PubMed]
3. Phulari, B.S. *An Atlas on Cephalometric Landmarks*; Jaypee Brothers Medical Publishers: New Delhi, India, 2013. [CrossRef]
4. Wang, C.; Huang, C.; Hsieh, M.; Li, C.; Chang, S.; Li, W.; Vandaele, R.; Marée, R.; Jodogne, S.; Chen, P.G.C.; et al. Evaluation and Comparison of Anatomical Landmark Detection Methods for Cephalometric X-Ray Images: A Grand Challenge. *IEEE Trans. Med. Imaging* **2015**, *34*, 1890–1900. [CrossRef] [PubMed]
5. Wang, C.; Huang, C.; Lee, J.; Li, C.; Chang, S.; Siao, M.; Lai, T.; Ibragimov, B.; Vrtovec, T.; Ronneberger, O.; et al. A benchmark for comparison of dental radiography analysis algorithms. *Med. Imag. Analy.* **2016**, *31*, 63–76. [CrossRef] [PubMed]
6. Lindner, C.; Wang, C.W.; Huang, C.T.; Li, C.H.; Chang, S.W.; Cootes, T. Fully automatic system for accurate localisation and analysis of cephalometric landmarks in lateral cephalograms. *Sci. Rep.* **2016**, *6*, 33581. [CrossRef] [PubMed]
7. Ibragimov, B.; Likar, B.; Pernuš, F.; Vrtovec, T. Shape representation for efficient landmark-based segmentation in 3-D. *IEEE Trans. Pattern. Analy. Mach. Intel.* **2014**, *33*, 861–874. [CrossRef] [PubMed]
8. Chen, R.; Ma, Y.; Chen, N.; Lee, D.; Wang, W. Cephalometric landmark detection by attentive feature pyramid fusion and regression-voting. In Proceedings of the International Conference on Medical Image Computing and Computer-Assisted Intervention, Shenzhen, China, 13–17 October 2019; Springer: Berlin/Heidelberg, Germany, 2019; pp. 873–881.
9. Li, W.; Lu, Y.; Zheng, K.; Liao, H.; Lin, C.; Luo, J.; Cheng, C.; Xiao, J.; Lu, L.; Kuo, C.; et al. Structured landmark detection via topology-adapting deep graph learning. In Proceedings of the Computer Vision–ECCV 2020: 16th European Conference, Glasgow, UK, 23–28 August 2020; Proceedings, Part IX, LNCS 12354, Glasgow, UKm 23–28 August 2020; Springer: Berlin/Heidelberg, Germany, 2020; pp. 266–283. [CrossRef]
10. Payer, C.; Štern, D.; Bischof, H.; Urschler, M. Integrating spatial configuration into heatmap regression based CNNs for landmark localization. *Med. Imag. Analy.* **2019**, *54*, 207–219. [CrossRef] [PubMed]
11. Song, Y.; Qiao, X.; Iwamoto, Y.; Chen, Y.W. Automatic Cephalometric Landmark Detection on X-ray Images Using a Deep-Learning Method. *Appl. Sci.* **2020**, *10*, 2547. [CrossRef]
12. Goodfellow, I.; Bengio, Y.; Courville, A. *Deep Learning*; MIT Press: Cambridge, MA, USA, USA, 2017.
13. Kingma, D.P.; Ba, J.L. Adam: A method for stochastic optimization. In Proceedings of the International Conference on Learning Representations, San Diego, CA, USA, 7–9 May 2015; Volume abs/1412.6980.

14. Conover, W. *Practical Nonparametric Statistics*, 3rd ed.; Wiley Series in Probability and Statistics; John Wiley & Sons: New York, NY, USA, 1999.
15. DeTone, D.; Malisiewicz, T.; Rabinovich, A. Deep Image Homography Estimation. *CoRR* **2016**. Available online: http://xxx.lanl.gov/abs/1606.03798 (accessed on 13 April 2022).
16. Goodall, C. Procrustes methods in the statistical analysis of shape. *J. Royal Stat. Soc. B* **1991**, *53*, 285–339. [CrossRef]
17. Rohlf, F.; Slice, D. Extensions of the Procrustes Method for the Optimal Superimposition of Landmarks. *Syst. Biol.* **1990**, *39*, 40–59. [CrossRef]
18. Cootes, T. *An Introduction to Active Shape Models*; Oxford University Press: Oxford, UK, 2000.

applied sciences

MDPI

Article

A Dermoscopic Inspired System for Localization and Malignancy Classification of Melanocytic Lesions

Sameena Pathan [1], Tanweer Ali [2,*], Shweta Vincent [3], Yashwanth Nanjappa [2], Rajiv Mohan David [2,*] and Om Prakash Kumar [2]

1 Department of Information and Communication Technology, Manipal Institute of Technology, Manipal Academy of Higher Education, Manipal 576104, India; sameena.bp@manipal.edu
2 Department of Electronics and Communication Engineering, Manipal Institute of Technology, Manipal Academy of Higher Education, Manipal 576104, India; yashwanth.n@manipal.edu (Y.N.); omprakash.kumar@manipal.edu (O.P.K.)
3 Department of Mechatronics, Manipal Institute of Technology, Manipal Academy of Higher Education, Manipal 576104, India; shweta.vincent@manipal.edu
* Correspondence: tanweer.ali@manipal.edu (T.A.); rajiv.md@manipal.edu (R.M.D.)

Abstract: This study aims at developing a clinically oriented automated diagnostic tool for distinguishing malignant melanocytic lesions from benign melanocytic nevi in diverse image databases. Due to the presence of artifacts, smooth lesion boundaries, and subtlety in diagnostic features, the accuracy of such systems gets hampered. Thus, the proposed framework improves the accuracy of melanoma detection by combining the clinical aspects of dermoscopy. Two methods have been adopted for achieving the aforementioned objective. Firstly, artifact removal and lesion localization are performed. In the second step, various clinically significant features such as shape, color, texture, and pigment network are detected. Features are further reduced by checking their individual significance (i.e., hypothesis testing). These reduced feature vectors are then classified using SVM classifier. Features specific to the domain have been used for this design as opposed to features of the abstract images. The domain knowledge of an expert gets enhanced by this methodology. The proposed approach is implemented on a multi-source dataset (PH2 + ISBI 2016 and 2017) of 515 annotated images, thereby resulting in sensitivity, specificity and accuracy of 83.8%, 88.3%, and 86%, respectively. The experimental results are promising, and can be applied to detect asymmetry, pigment network, colors, and texture of the lesions.

Keywords: color; classification; dermoscopy; hair shafts; melanoma; segmentation; shape; texture; pigment network

Citation: Pathan, S.; Ali, T.; Vincent, S.; Nanjappa, Y.; David, R.M.; Kumar, O.P. A Dermoscopic Inspired System for Localization and Malignancy Classification of Melanocytic Lesions. *Appl. Sci.* **2022**, *12*, 4243. https://doi.org/10.3390/app12094243

Academic Editors: Ioannis A. Kakadiaris, Michalis Vrigkas and Christophoros Nikou

Received: 23 February 2022
Accepted: 13 April 2022
Published: 22 April 2022

1. Introduction

1.1. Motivation

Melanoma is one of the worst forms of skin cancer resulting in increased morbidity and huge medical expenditure almost up to $3.3 billion [1]. Although, simple observation aids in detection of the changes in the melanocytic nevi, the deadly disease can spread to other parts of the body by metastasizing. Skin tumor thickness mainly determines the spread of the disease. Melanoma prognosis is inversely proportional to the tumor thickness, since the survival rate relies on the tumor thickness. The greater the thickness, the lesser the survival rate. However, to measure the spread and thickness of the tumor biopsy is required, which is a painful experience to the patient. Additionally, careful observation of the melanoma characteristics can be performed, as the lesion is visible on the skin. However, it is further liable to metastasize and spread to lymph nodes thus incrementing the level of malignancy. According to the Fitzpatrick Skin classification there are six skin types [2]. The type I and type II are more prone to melanoma compared to the other skin types. Melanocytes produce a pigment termed as melanin, which gives the natural color to the skin. There are two kinds of melanin, eumelanin and pheomelanin present in the dark skinned and lighter skinned population respectively. Eumelanin is insoluble and thus the skin darkening effects produced by eumelanin last relatively longer compared to the skin-reddening effects produced by pheomelanin. A Dermoscope aids the dermatologists in the primitive analysis of melanocytic skin lesions [2]. Owing to a dearth of experience and differences in visual perceptions, the prognosis of melanoma is still subjective, in spite of the availability of well-established medical methods. This fosters the need for an objective evaluation tool. Computer Aided Diagnostic (CAD), tools were introduced for dermoscopic images to provide quantitative and target assessment of the skin lesion to help clinicians in demonstrative and prognostic undertakings. Due to the inter and intra-observer variabilities, determination of melanoma is innately subjective. Thus, a CAD tool institutively eliminates the subjectivity in the diagnosis and prognosis of melanoma, and aids in early detection of melanoma in situ, thereby improving the accuracy of detection and reducing the mortality rate. This paper describes a clinical framework that can significantly identify the lesion properties and provide a diagnosis. The system incorporates the knowledge of an experienced dermatologist to co-relate the features extracted with their histopathological relevance. Additionally, it further incorporates certain statistical features to achieve promising results.

1.2. Related Work

The literature reports numerous studies for designing CAD systems used for the diagnosis of melanoma. Based on the features used for the prediction of the lesion, approaches for the diagnosis of melanoma have been broadly classified into three types: (i) methods inspired by the dermoscopic criteria (ABCD), that take into account the global and local lesion features [3–5]; (ii) methods based on the characteristic properties of images [6–8]; and (iii) combination of the aforementioned methods. These methods can be used to either develop a cumulative score [9] or can be used to develop trained models that make predictions. The proposed approach in this paper, belongs to the third category. Nevertheless, the literature also indicates a few approaches that combine the two categories [10,11]. Celebi et al. [10] used color, texture and presence of blue-white veil for classification of skin lesions. However, the lesions were segmented manually, to separate issue of feature extraction from automated border detection. The smooth boundary between the lesion and surrounding skin, poses difficulties in automated border detection. Abuzaghleh et al. [11] classified the lesions using abstract image features and two dermoscopic features. However, the method is computationally complex due extraction of large feature sets. The method proposed in [5] concentrates mainly on the clinical aspects of color in dermoscopic images and texture features, while missing out the important role of shape features. Several studies also report the use of complex deep learning architecture for segmentation and classification [12,13]. However, these methods need to tackle the vanishing gradient and degradation problem.

1.3. Problem Statement

Based on the literature, a clinical framework for the diagnosis of melanoma should encompass the requirements mentioned below.

1. Provide automated localization of the lesions.
2. The features extracted need to hold clinical significance.
3. Result in a balance between sensitivity and specificity for distinguishing the lesion classes.

The aforementioned issues are addressed in this work.

1.4. Contributions

The proposed system describes the development of a clinically inspired framework for diagnosis of melanocytic lesions, such that the system is informative from the perspective of a dermatologist. In contrast to the methods proposed in literature [3–14], the proposed system considers lesion specific properties in order to distinguish benign and malignant melanocytic lesions. Additionally, rather than extracting abstract image features, algorithms are developed for the extraction of melanocytic specific features.

1.5. Summary

The manuscript has been organized in the following manner: section two provides the technical depths pertaining to the methodologies developed for the extraction of hair shafts, segmentation of lesion masks, dermoscopic feature extraction and classification. The section three reports the quantitative results obtained. The manuscript concludes with the discussion and conclusion section that provide key aspects with respect to the methodologies developed and future research perspectives.

2. Methods

This section describes the proposed framework. The proposed framework's overview is illustrated in Figure 1.

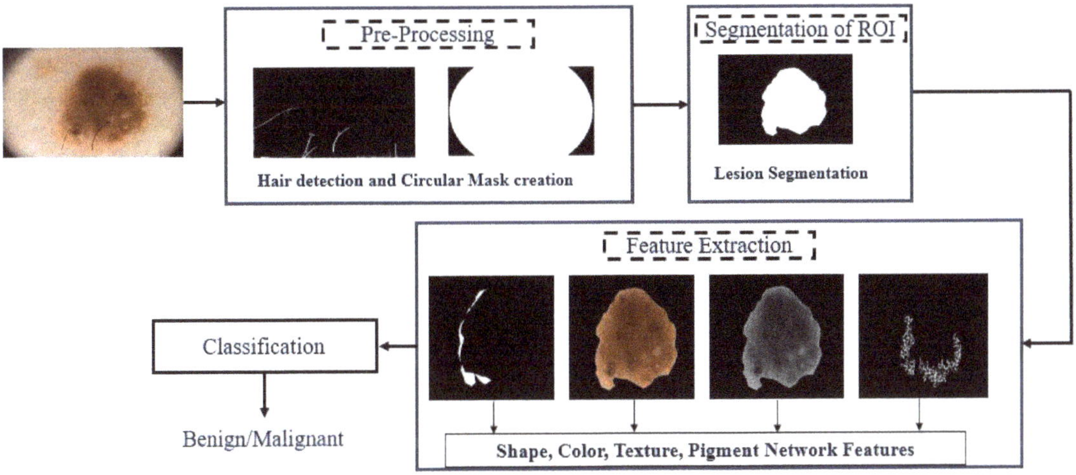

Figure 1. Overview of the proposed system.

Initially, preprocessing of the dermoscopic images is performed for eliminating artifacts viz., dark corners, ruler markings, hair and dark frames. For the elimination of dark corners, masks of circular shape are created with a radius and centroid co-ordinates computed as given in (1).

$$Radius = \frac{\max(r,c)}{2},\ Centroid = \left(\frac{r}{2}, \frac{c}{2}\right) \tag{1}$$

where, r, c correspond to the rows and columns of the dermoscopic image. The Figure 2 illustrates the circular mask created for the corresponding dermoscopic image Figure 2A.

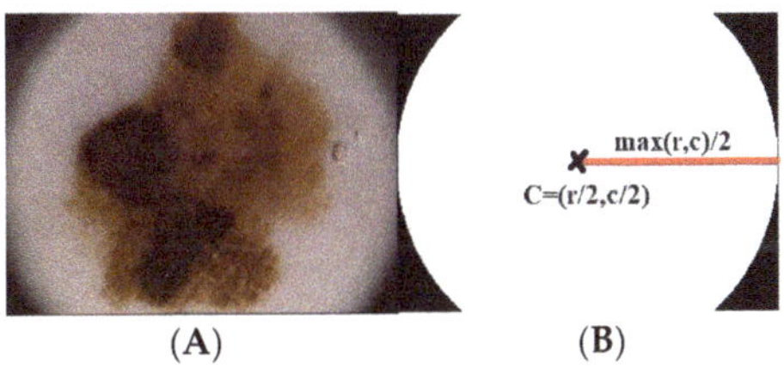

Figure 2. Mask created (**A**) Dermoscopic Image (**B**) Mask for image (**A**).

The hair masks are multiplied with the initial contour prior to curve deformation to eliminate the dark corners.

2.1. Detection and Removal of Hair

Numerous techniques have been proposed for the detection and exclusion of hair [11,15–18]. These techniques have been designed assuming that hair color is much darker than the skin and lesion. Additionally, the properties of dermoscopic hair shaft were not considered to detect the dermoscopic hair. Owing to the localization of melanin in the upper and lower epidermis, most skin lesions are either brown or black in color. Therefore, consideration of color variation between hair and surrounding skin could be erroneous leading to an overlap between the attributes of the lesion and hair. This signifies that, for a hair detection algorithm, a need exists for the inclusion of attributes specific to dermoscopic hair.

Geometric deformable models and their success depends on the initial conditions and speed function evolution. As the color of melanin is dependent on the extent of localization in the skin, the attribute of color is vital in the creation of this framework [19–22]. Therefore, the approach of segmentation has been adopted in this study by giving consideration to the chroma component as opposed to the RGB channels used in conventional systems. The Figure 3 illustrates the results of dermoscopic hair detection, for the corresponding dermoscopic images of Figure 3A and Figure 3B, the hair masks obtained are Figure 3C and Figure 3D, and the Figure 3E and Figure 3F, corresponds to dermoscopic images after hair inpainting.

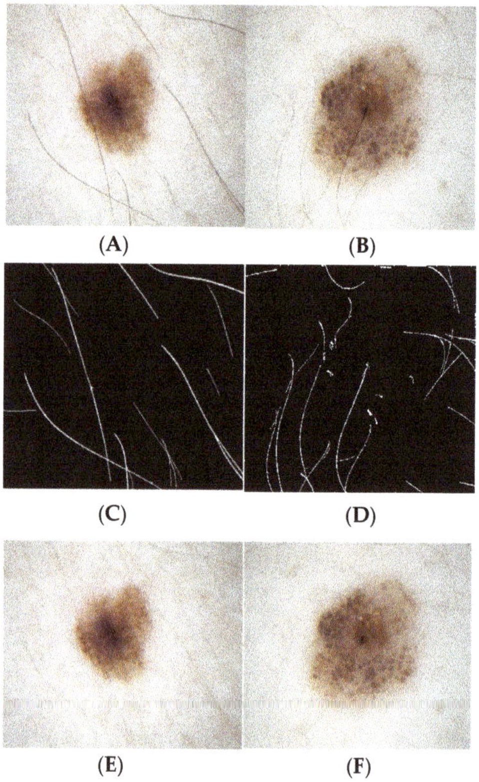

Figure 3. Hair shaft detection and exclusion method (**A**,**B**) dermoscopic images, (**C**,**D**) hair shafts detected, and (**E**,**F**) dermoscopic images after inpainting.

The Figure 4 illustrates the process of segmentation. Figure 4D illustrates the segmented border obtained by the proposed approach and the boundary of the ground truth.

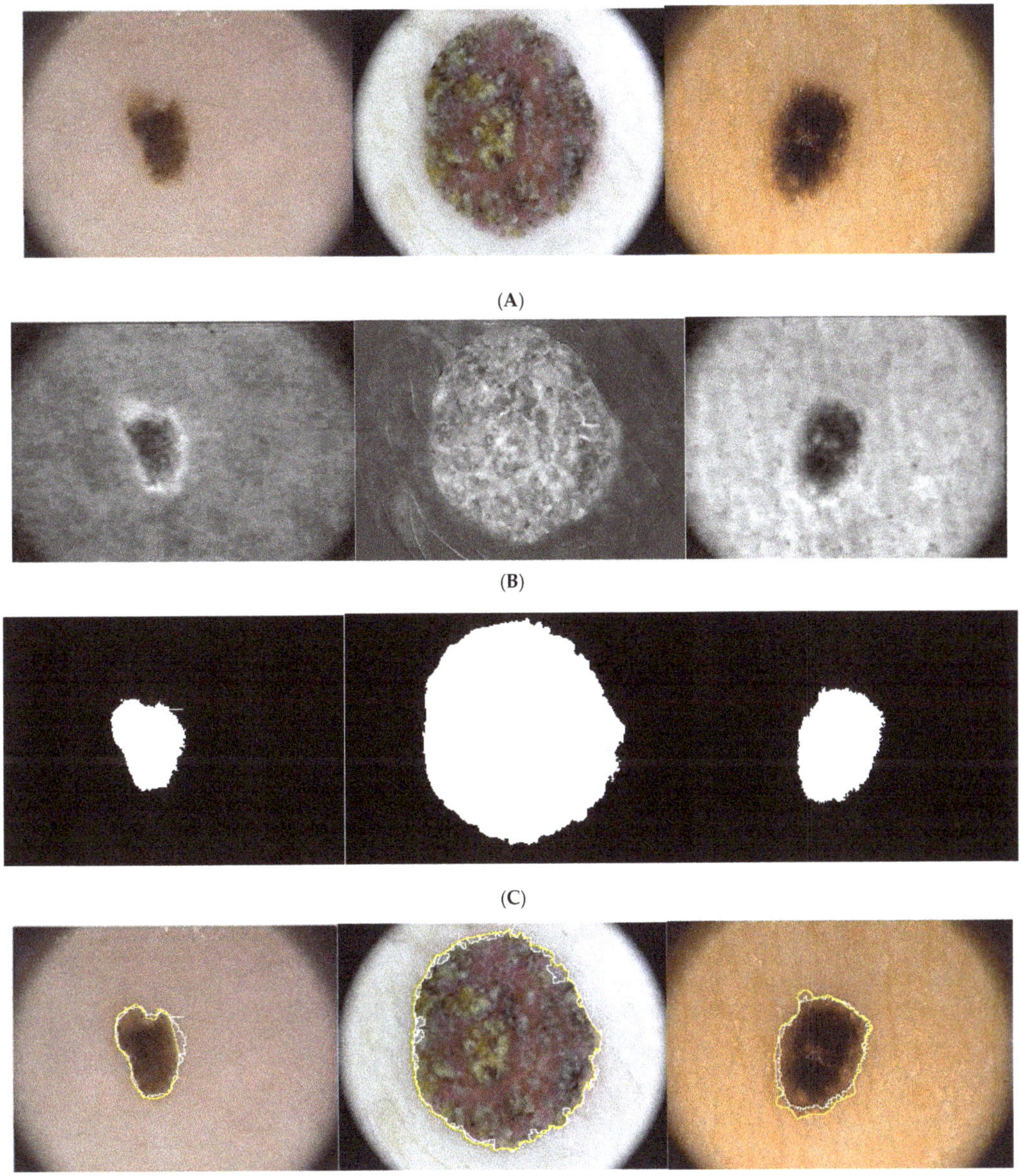

(A)

(B)

(C)

(D)

Figure 4. Illustration of the proposed segmentation approach: (**A**) original Images, (**B**) chroma component, (**C**) segmented images, (**D**) boundary of ground truth and segmented region overlapped on the original images (yellow corresponds to ground truth, white corresponds to segmented output.

2.2. Extraction and Classification of Features

2.2.1. Color Features

The role of color in dermoscopy is inevitable. The most important chromophore of the skin is melanin. Lesions which are benign exhibit one or two colors. Since the malignant lesions are localized within the deeper structures of the skin, they tend to exhibit three or more colors. To study the color properties of the lesions, six groups of features were computed namely, color asymmetry, color similarity index, color entropy and statistical color features (i.e., color co-relation coefficient, principal component analysis and color entropy). The statistical color features are derived specifically for two categories: (i) region of interest (ROI) (ii) between ROI and non-ROI (NROI). The color features are delineated as follows:

The color asymmetry and color similarity index draw their inspiration from the ABCD rule of dermoscopy. The color asymmetry is quantified by the difference between the opposite halves of the lesion along x-axis (Cx_1, Cx_2) and y-axis (Cy_1, Cy_2). The perceived color difference ΔE is calculated in the CIE L^*a^*b color space as given in (9). The four halves are divided as indicated in Figure 5. Correspondingly, four asymmetry indices are computed as given in (2).

$$\begin{aligned} Cx_1 &= \Delta E_1 - \Delta E_3 \\ Cx_z &= \Delta E_2 - \Delta E_4 \\ Cy_1 &= \Delta E_1 - \Delta E_2 \\ Cy_2 &= \Delta E_3 - \Delta E_4 \end{aligned} \quad (2)$$

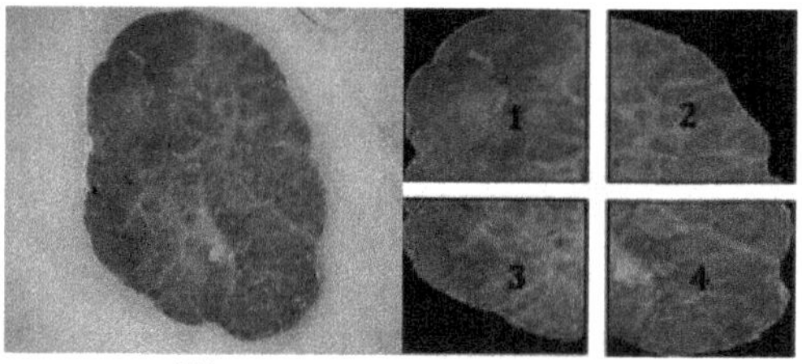

(A) (B)

Figure 5. Color asymmetry calculation: (**A**) dermoscopic image; (**B**) the four halves of ROI.

A set of four color asymmetry indices were computed.

Color similarity index indicates the presence of six suspicious colors in a lesion (light-brown (LB), dark-brown (DB), black (K), white (W), red (R) and blue-gray (BG)). The color similarity index is computed by computation of the Euclidean distance between the lesion RGB values and corresponding six suspicious colors. The color of the lesion pixels is used to determine the color similarity index and hence the lesion masks are used for computing the color similarity index. The color similarity index and Euclidean distance are inversely related as given in (3).

$$Euclidean\ Distance \propto \frac{1}{Color\ Similarity} \quad (3)$$

The Algorithm 1 summarizes the steps used for calculating the color similarity score. The threshold Th is determined from the RGB values of two opposite colors [9]. The two opposite colors are black and white. Further, a score of 1 is assigned if more than two percent of the lesion area has suspicious color.

Algorithm 1. Color Similarity Index Calculation

$Input: \ I(x,y) = [R(x,y), G(x,y), B(x,y)]$
$Output: \ Score$
$Lesion\ ROI: \ M(x,y) \in I(x,y)$
$S = [RS, GS, BS] \ : \ suspicious\ color$
$output \ : \ score$
$score = 0$
$Th = 0.5\,[\,(Rw - Rk)2 + (Gw - Gk)2 + (Bw - Bk)2\,]0.5$
$C = 0$
$for\ each\ pixel\ (RM\_i, GM\_i, BM\_i)\ in\ M\ do$
$D = [\,(RM\_i - RS)2 + (GM\_i - GS)2 + (BM\_i - BS)2\,]0.5$
$if\ D <= Th$
$C++$
$if\ C >= 0.2 * (sum(M))$
$score = 1$
$return\ score$

The color variance is computed considering the red, green, blue and gray-scale values for the lesion along with the entire image, thus resulting in computation of eight features (VR, VG, VB, and VK). The degree of randomness quantified by color entropy (E) is computed similarly for the red and blue values of the lesion and entire image, leading to computation of four features (ER and EB). The correlation co-efficient signifying the direction and degree of closeness of intra and inter linear relations between the RGB channels and grayscale images resulted in computation of twelve features (CRG, CGB, CBR, CRK, CGK, and CBK). Along with these, the lesion RGB values are projected on the three principal components (PC). Therefore, a total of 37 color features were computed.

2.2.2. Texture Features

According to the opinion of expert dermatologists malignant melanocytic lesions are characterized by coarse texture with a contrast which is inhomogeneous and irregular patterns. Since Tamura's et al. [23] texture features are based on the visual perception, a set of three texture features, namely coarseness (T1), contrast (T2) and directionality (T3) were computed. A larger value of coarseness specifies a greater degree of roughness. Coarseness is calculated as the average of the best size that gives the maximum difference between the non-overlapping neighborhoods in horizontal and vertical directions. Directionality is computed by taking the gradient of the neighboring pixels given by (4).

$$\Delta G = \frac{|\Delta H| + |\Delta V|}{2} \tag{4}$$

where, ΔG is the edge strength, ($|\Delta H|$) and ($|\Delta V|$) indicate the horizontal and vertical change in direction. Further, the contrast is calculated as the statistical distribution of pixel gray values.

2.2.3. Shape Features

The computation of the shape symmetry index is performed using the lesion mark. The lesion centroid is positioned at the centroid of the image by using the difference in centroid positions method showcased in (5). This is carried out since the lesions are not aligned with the center of the image.

$$\Delta\{x,y\} = \{(C_I(x) - C_L(x)), (C_I(y) - C_L(y))\} \tag{5}$$

$C_I(x,y)$ and $C_L(x,y)$ indicate the image and lesion centroids respectively. The image is divided into two halves with respect to x-axis and y-axis as illustrated in Figure 6 to determine the asymmetry along x and y axis. The maximum possible asymmetry index of the lesion AI is given in (6).

$$AI = max\{A_x = x_1 \hat{} x_2\, , A_y = y_1 \hat{} y_2\, , \}/2 \quad (6)$$

where, x_1 and x_2 are the two halves with respect to the x-axis and y_1 and y_2 are the two halves with respect to the y-axis. Since, asymmetry quantifies malignancy the proposed method takes into account the maximum possible asymmetry to minimize the classification error.

(A) **(B)** **(C)** **(D)**

Figure 6. AI calculation: (**A**) dermoscopic image, (**B**) left half of the image, (**C**) right half of the image, (**D**) asymmetric region over y-axis.

A multi-scale method termed as fractal dimension (FD) is used to quantify border irregularity. It is computed by dividing the image in small grids of size $r \times r$ as given in (7) [9].

$$log\frac{1}{N(r)} = fd \times \log(r) - \log(\lambda) \quad (7)$$

$N(r)$ gives the contour length, λ indicates the scaling constant. The circularity of the lesion is measured using the metric of Compactness (8) [24].

$$CI = \frac{{P_L}^2}{4\pi A_L} \quad (8)$$

P_L indicates the perimeter of the lesion and A_L indicates the area of the lesion.

2.2.4. Detection of the Pigment Network

Pigment network is a honeycomb-like structure characterized by linear strokes and directional shapes. The presence of pigment network histopathologically, indicates the melanin presence in keratinocytes and melanocytes at the dermal and epidermal junction [25]. Additionally, the lines in the pigment network have diverse orientation. Thus, the proposed approach for detection of pigment network is similar to the hair detection method proposed in Section 2.1 with a few additional steps and change in Gabor parameter f. The green channel is used for processing due to relatively greater contrast. A second order derivative Laplacian operator [3 × 3 mask] is used for enhancing the finer details in the image after median filtering. The enhanced image is convolved with the 2D Gabor filter defined in (3). The empirically determined values for σ_x and σ_y are same. However, the value of thickness parameter t is set to 3.3, since the lines of the pigment network are comparably thinner than the hair shafts.

For post processing the Adaptive Histogram Equalization (AHE) is followed by determination of the threshold to extract the pigment network efficiently. The threshold is computed by fitting a fourth degree polynomial as given in (9) to the contrast enhanced image.

$$n = mc^4 + ac^3 + bc^2 + dc + e \quad (9)$$

$m,\ a, b, d$ and e are three different points to fit a curve with c distinct co-ordinates. The pigment network mask serves to calculate the five distinct features $(f_1, f_2, f_3, f_4, f_5)$ as given in [25]. The Figure 6 illustrates the pigment network detection process. A comparison of the proposed pigment network detection method with the method proposed by Barata et al. [15] is illustrated in Figure 7. It can be observed by from Figure 8B,C that, the proposed method accurately identifies honeycomb-like pigment network structures in comparison to the method in [15]. An overview of the features extracted is summarized in Table 1.

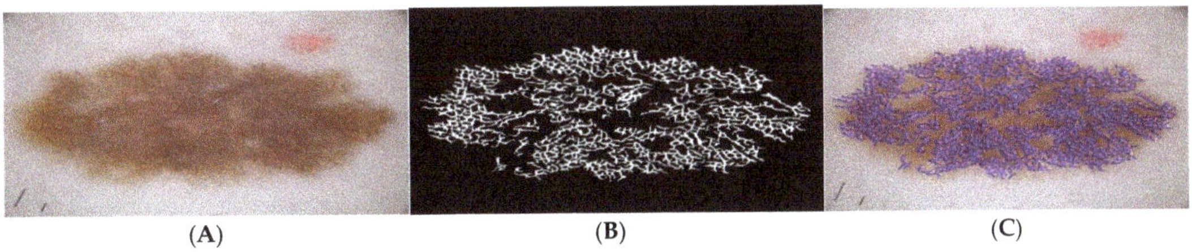

Figure 7. Detection of pigment network: (**A**) dermoscopic image, (**B**) pigment network mask detected, (**C**) corresponding mask ((**A**) overlaid on (**B**)).

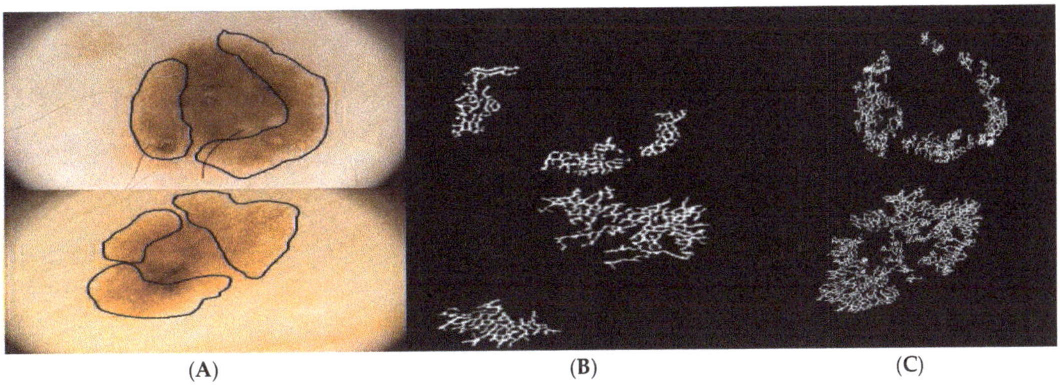

Figure 8. Comparison of pigment network detection: (**A**) dermoscopic images with pigment network marked, (**B**) pigment network masks detected by Barata et al. [15], (**C**) pigment network masks detected by the proposed method.

Table 1. Overview of the features extracted.

Feature Type	Description (Number)
Shape	Shape Asymmetry Index (1), Compactness Index (1), and Fractal Dimensions (1)
Color	Color Asymmetry Index (4), Color similarity score (6), Color variation (8), color entropy (4), color co-relation (12), and PCA (3),
Texture	Coarseness (1), Contrast (1), and Directionality (1)
Dermoscopic Structure	Pigment Network (5)

2.2.5. Classification and Diagnosis

The features selected from the groups of f_{shape}, f_{color}, $f_{texture}$, f_{PN} are concatenated to benefit from the complimentary information captured by the feature types. The classification of the observations into two classes classifier (benign and malignant) C yielding the largest probability $P(G)$ which is performed using a probabilistic SVM, as given in (10).

$$C = \max\left(P_{shape}(G),\ P_{color}(G), P_{texture}(G), P_{PN}(G),\right) \quad (10)$$

$P_{shape}(G)$, $P_{color}(G)$, $P_{texture}(G)$, $P_{PN}(G)$ indicates the cumulative probabilities of shape, color, texture and pigment network features. The features are concatenated and used to train a single SVM classifier. Platt's method is used for the computation of posterior probabilities [26,27]. Linear kernel is used to map the scores.

3. Results

3.1. Dataset and Evaluation Metrics

A multi-source dataset of 515 images which was taken from PH^2 [28], ISBI 2016 and ISBI 2017 [29,30], has been used for experimentation in this article. The dataset consists of 304 benign and 211 melanoma lesions with annotated ground truths. In the initial stage, pre-processing of the images is performed for the removal of dermoscopic hair and dark corners. The algorithms have been implemented using MATLAB 2016®. Three metrics viz., Sensitivity (SE), Specificity (SP) and Accuracy (ACC) have been used for the detection of ROI and lesion classification. In addition to this the overlap error (between the ground truth and segmented mask) is also computed for evaluation of lesion segmentation. The null hypothesis testing has been performed using the Wilcoxon Rank Sum statistics which is a non-parametric test. The null hypothesis is stated below:

H_0. *The extracted features for benign and malignant lesions have equal medians.*

The testing of the null hypothesis against the alternative hypothesis is performed. The alternate hypothesis states that the features extracted do not have equal medians and hence are statistically significant at 5% significance level. Among the 48 p-values, 34 p-values satisfied the alternative hypothesis and hence were used for classification. A hold-out set of 25% is used for testing. The classification metrics are computed by repeating the training and test procedures ten times by stratifying the training and test sets.

3.2. Evaluation of Hair Detection and Lesion Segmentation: Results

Hair detection and exclusion is performed prior to lesion segmentation to eliminate the artifacts thereby increasing the accuracy of segmentation. A positive effect of pre-processing (hair detection + black frame removal) on the segmentation accuracy for the combined dataset can be observed from the Figure 9. The overlap error after applying hair detection algorithm prior to segmentation was 0.07, and the overlap error obtained without applying hair detection algorithm prior to segmentation was 0.15. This proves that hair detection improved the accuracy of segmentation.

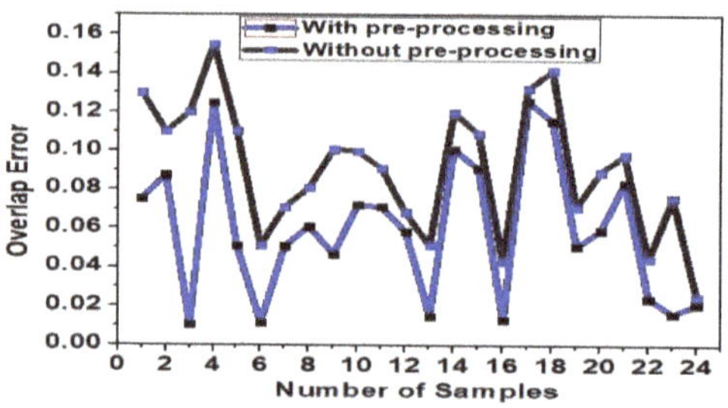

Figure 9. Effect of pre-processing on segmentation accuracy.

The proposed segmentation method resulted in sensitivity, specificity, accuracy and overlap error of 92.5%, 96.7%, 95.7%, and 8.2%, respectively, for the combined datasets. The overlap segmentation error for modified Chan-Vese (proposed method) and Chan-Vese for combined dataset is illustrated in Figure 10. The Overlap Error (OE) is calculated as given in (11).

$$OE = \frac{Area(G \oplus S)}{Area\ (G)} \tag{11}$$

where G is the ground truth and S is the segmented binary image.

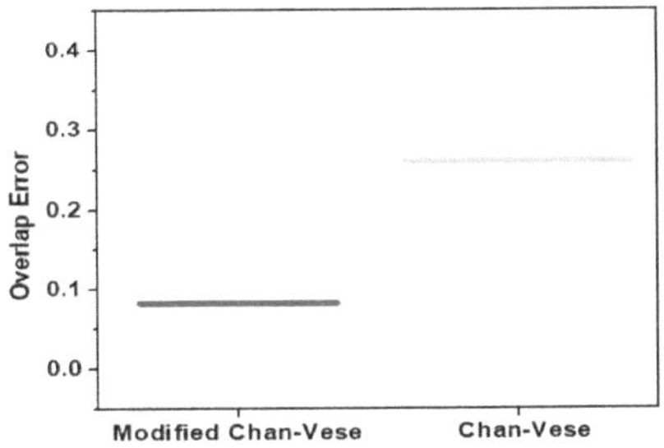

Figure 10. Overlap segmentation error for Modified Chan-Vese and Chan-Vese Algorithm.

3.3. *Evaluation of Features Extracted and Lesion Classification: Results*

Various experiments were conducted to deduce the classifier with a major goal to assess the best subset of features and compare the performance of training the SVM with single set of features against training with concatenation of features. The Figure 11 illustrates the plot of features for combined dataset versus the p-values. A good score has a p-values less than or equal to 0.05 ($p \leq 0.05$) whereas $p > 0.05$ is considered a bad score. It is seen from Figure 11, that the shape and pigment network features have good scores. However, the performance scores for color asymmetry index (x_2, x_3, x_4), indicates comparably lower scores. Insignificant p-values for color similarity index for colors red, light brown and dark brown can be observed. This justifies the fact that presence of red is due to vascularization of blood vessels, irrespective of the lesion class and shades of brown are common to both the lesion types (benign and malignant). Regarding texture features, T2 (contrast) performed comparably poor then T1 and T3. Similarly, the role of statistical color features for ROI and NROI can be interpreted from Figure 11. The Table 2 provides the mean and standard deviation values of the features with significant p-values for combined datasets.

Table 2. Mean and standard deviation values of the features with significant p-values.

F	Mean	SD	F	Mean	SD
AI	0.69	0.94	VRI	1509.015	1368.13
CI	2.63	3.21	VGI	1834.916	1303.59
FD	26.31	9.30	PC1	2910.31	1911.63
T1	39.80	17.70	PC2	116.10	100.96
T3	13.42	12.77	PC3	11.62	7.95
Cx1	13.57	12.89	ER	6.54	0.65
W	0.10	0.31	EB	6.62	0.44
K	0.24	0.42	ERI	6.19	0.75
BG	0.94	0.21	EBI	6.80	0.47
CRG	0.01	0.14	F1	7361.83	22,929.7
CGB	0.94	0.05	F2	0.08	0.17
CBR	0.95	0.09	F3	0.52	0.39
CRK	0.85	0.10	F4	0.06	0.43
CGK	0.99	0.05	F5	0.14	0.16
CBK	0.94	0.06			
CRGI	0.93	0.05			

Table 2. *Cont.*

F	Mean	SD	F	Mean	SD
CBRI	0.86	0.10			
VR	1032.02	832.46			
VG	1032.52	702.47			
VK	974.10	652.87			

Note: F—Feature, SD—Standard deviation, AI—Asymmetry Index, CI—Compactness Index, FD—Fractal dimensions, T1—Coarseness, T3-Directionality, Cx1—Color symmetry index, W—white, K—Black, BG—Blue Gray, CRG—Color Variation between Red and Green, CGB—Color Variation between Green and Blue, CBR—Color Variation between Blue and Red, CRK—Color Variation between Red and Grey, CGK—Color Variation between Green and Grey, CBK—Color Variation between Blue and Grey, CRGI—Color Variation between Red and Green for entire image, CBRI—Color Variation between Blue and Red for entire image VR—Color Variation for Red, VG—Color Variation for Green, VK—Color Variation for grey, VRI—Color Variation for Red for entire image, VGI—Color Variation for Green for entire image, PC1, PC2, PC3—Three Principal components, ER-Entropy for red, EB—Entropy for Blue, ERI—Entropy for Red for entire image, EBI—Entropy for Blue for entire image, F1–F5—Pigment network features.

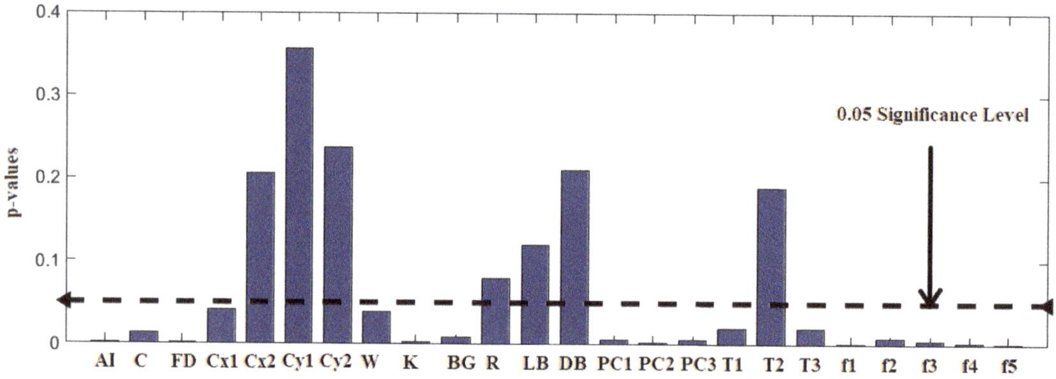

(**A**)

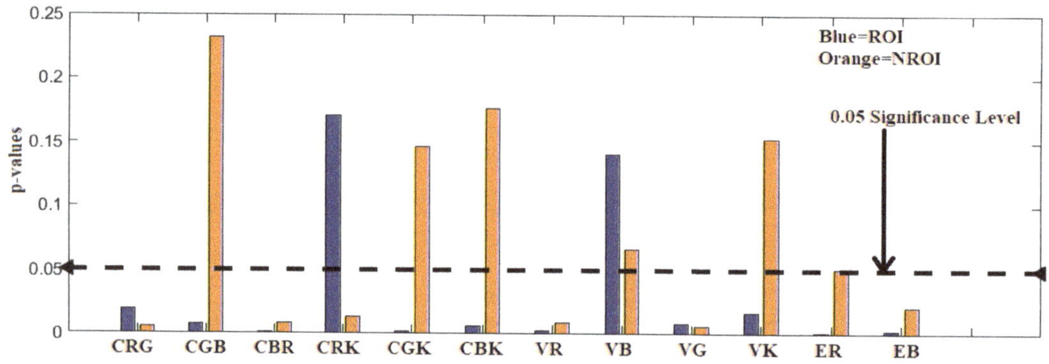

(**B**)

Figure 11. Plot of features extracted versus the *p*-values: (**A**) lesion specific features, (**B**) statistical color features (CRG, CGB, CBR, CGK, and CBK indicate correlation between red (R), green (G), blue (B), gray values (K)), V indicates color variance, and E indicates entropy).

Based on the *p*-values the feature set was reduced from 48 to 34 features. The role of single and combined features in lesion diagnosis is given in Table 3. It can be inferred from the Table 3 that, the role of color features is significantly pre-dominant for lesion diagnosis, followed by pigment network features. Interestingly, the best overall results were obtained for a combination of the features. Table 4, shows the results of applying the proposed framework on diverse datasets, correspondingly the plot of ROC for the same is illustrated in Figure 12. It can be observed from the Table 4, acceptable results were obtained for diverse datasets, with PH^2 dataset yielding the greatest accuracy. The generalization ability of the features extracted is tested by training on ISBI dataset and testing on PH^2 dataset and vice-versa, the results are depicted in Table 5.

Table 3. Contribution of features for lesion diagnosis (PH^2 Dataset).

Set-Up	SE (%)	SP (%)	ACC (%)
F_{shape}	90.4	82.7	83.5
F_{color}	88.8	92.8	91.9
$F_{texture}$	78.7	85.4	84.4
F_{PN}	88.7	84.2	86.5
$F_{combined}$	95.6	95.1	95.3

Table 4. Classifier Performance for different datasets.

Dataset	SE (%)	SP (%)	ACC (%)
PH^2	95.6	95.1	95.3
ISBI 2016 + 2017	83.4	93.7	85.4
Combined	83.8	88.3	86

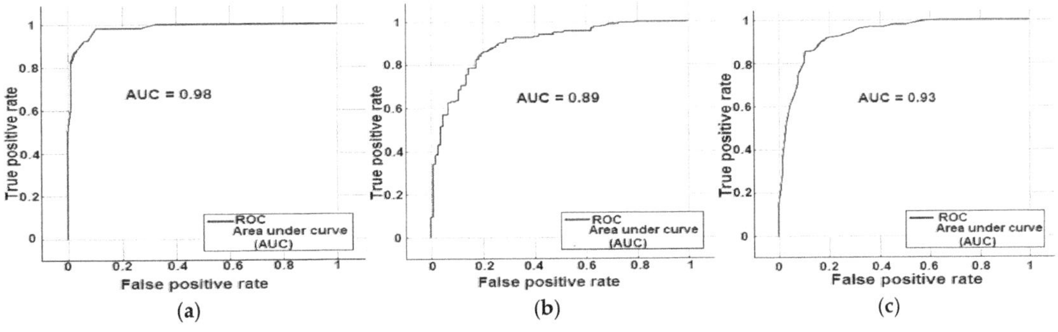

Figure 12. ROC curves (**a**) For PH^2 data (**b**) For ISBI data (**c**) For Combined datasets.

Table 5. Classifier performance depicting the classifier generalization ability.

Dataset	SE (%)	SP (%)	ACC (%)
ISBI on PH^2	80.5	81.5	80.7
PH^2 on ISBI	90	75	81.2

4. Discussion

The major objective of this study is the development of an automated computer aided melanoma diagnostic system using clinical aspects of dermoscopy on a diverse dataset. In this regard, the diagnosis system was built using a sequence of algorithms for pre-processing, ROI extraction, feature extraction and classification. Since these steps are sequential in nature, the accuracy of classification mainly relies on the efficiency of the preceding steps. The hair detection algorithm considers the dermoscopic knowledge of hair shafts thereby neglecting a overlap of the attributes of lesion and hair. Such an algorithm prevents loss of lesion specific information and efficiently eliminates the light and dark hair that subsequently aids in improving the lesion segmentation accuracy. Border irregularity is a major indication of malignancy of melanocytic lesions. Thus, while localizing the ROI appropriate care has to be taken to prevent the loss of lesion border details. Geometric deformable models incorporating color information have provided promising segmentation accuracy even in the presence of background noise and poor contrast. The segmentation process was followed by extraction of a set of 48 features specific to shape, color, texture and pigment network from the segmented ROI's to facilitate identification of benign and malignant lesions. The non-parametric Wilcoxon Rank Sum statistics was used to obtain the p-values for the features extracted with the goal of finding the best features.

Of late, deep learning techniques have been extensively used in skin lesion classification [12,13,31]. In spite of the fact that these architectures have increase the accuracy of classification using large data for learning, the optimization of network parameters for reducing computational complexity is unexplored. A quantitative comparison of the proposed method and the state of art methods reported above may be tenuous due to the diversity of the datasets involved, however a comparative analysis of the studies carried out using the same datasets is given in Table 6. Sensitivity indicates the accurate rate of classification of melanoma lesions. Specificity indicates the accurate rate of classification of benign lesions, whereas accuracy gives a cumulative score of classification of benign and malignant lesions. In [5], the sensitivity was higher in contrast to specificity, since the main focus was on color feature of the lesions. An imbalance in sensitivity and specificity was obtained by Yu et al. [31], by employing a deep learning-based architecture. A methodological approach to detect pigmented skin lesions was proposed in [32]. Pennisi et al. [33], have used standard color, shape and texture feature for classification after applying Delaunay Triangulation based segmentation acc Nonetheless, the comparison provides us with relevant information about the significance of the proposed method. Nonetheless, the proposed method employs domain specific features, thereby improving the accuracy in classification of benign and malignant lesions. However, the study did not consider the thickness feature of the lesion due to lack of third dimensional image data and ground truth. The thickness feature would be an important parameter to rate the stage of malignancy once, the lesion malignancy has been detected by the classification model. Another limitation of the study if the processing time, since it approximately takes 90 s, on an average system of 8 GB RAM, and clock frequency of 1.60 GHz to provide the diagnosis once, the dermoscopic image is given as the input to the system The trained system can be employed in a clinical scenario, by using a dermoscopic based image capturing system, since a dermo scope would enhance the resolution of the lesions that would aid in better analysis, rather than a normal image capturing device.

Table 6. Comparative analysis of lesion classification methods with the state-of art.

Dataset	Ref.	SE (%)	SP (%)	ACC (%)
PH^2	Barata et al. [5]	**100**	88.2	-
	Pennisi et al. [33]	93.5	87.1	
	Proposed	95.6	**95.1**	**95.3**
ISBI 2016 + 2017	Yu et al. [31]	54.7	93.1	85
	Proposed	**83.4**	**93.7**	**85.4**

5. Conclusions

This paper presents the development of a clinically oriented framework for melanoma diagnosis. On the basis of the color characteristics of the lesion, the regions are segmented. It can be observed from the Table 4 that the role of color is evident in melanoma detection relative to other features. However, the color features could be severely affected due to variations in image acquisition modalities. Hence, while acquiring real-time images, appropriate illumination correction techniques should be employed to eliminate the effects of non-uniform illuminations.

The experimental results are promising and can be applied to detect asymmetry, pigment network, colors and texture of the lesions. Finally, the detected criteria are combined to develop a cumulative model which exhibits sensitivity, specificity and accuracy of 83.8%, 88.3%, and 86%, respectively.

Author Contributions: Conceptualization, S.P., T.A., S.V., Y.N., R.M.D. and O.P.K.; Formal analysis, S.P., T.A., R.M.D. and O.P.K.; Investigation, S.P. and S.V.; Methodology, S.P., Y.N. and O.P.K.; Software, S.P.; Visualization, T.A. and S.V.; Writing—original draft, S.P.; Writing—review & editing, T.A., Y.N., R.M.D. and O.P.K. All authors have read and agreed to the published version of the manuscript.

Funding: This research received no external funding.

Institutional Review Board Statement: Not applicable.

Informed Consent Statement: Not applicable.

Data Availability Statement: Not applicable.

Conflicts of Interest: The authors declare no conflict of interest.

References

1. The Medical Futurist. Amazing Technologies Changing the Future of Dermatology—The Medical Futurist. 2017. Available online: http://medicalfuturist.com/future-of-dermatology/ (accessed on 24 September 2017).
2. Pathan, S.; Prabhu, K.G.; Siddalingaswamy, P.C. Techniques and algorithms for computer aided diagnosis of pigmented skin lesions—A review. *Biomed. Signal Process. Control* **2018**, *39*, 237–262. [CrossRef]
3. Abbas, Q.; Celebi, M.E.; García, I.F. Skin tumor area extraction using an improved dynamic programming approach. *Ski. Res. Technol.* **2011**, *18*, 133–142. [CrossRef] [PubMed]
4. Abbas, Q.; Celebi, M.E.; Garcia, I.F.; Ahmad, W. Melanoma recognition framework based on expert definition of ABCD for dermoscopic images. *Ski. Res. Technol.* **2013**, *19*, e93–e102. [CrossRef] [PubMed]
5. Barata, C.; Celebi, M.E.; Marques, J. Development of a clinically oriented system for melanoma diagnosis. *Pattern Recognit.* **2017**, *69*, 270–285. [CrossRef]
6. Garnavi, R.; Aldeen, M.; Bailey, J. Computer-Aided Diagnosis of Melanoma Using Border- and Wavelet-Based Texture Analysis. *IEEE Trans. Inf. Technol. Biomed.* **2012**, *16*, 1239–1252. [CrossRef] [PubMed]
7. Kostopoulos, S.A.; Asvestas, P.A.; Kalatzis, I.K.; Sakellaropoulos, G.C.; Sakkis, T.H.; Cavouras, D.A.; Glotsos, D.T. Adaptable pattern recognition system for discriminating Melanocytic Nevi from Malignant Melanomas using plain photography images from different image databases. *Int. J. Med. Inform.* **2017**, *105*, 1–10. [CrossRef] [PubMed]
8. Ferris, L.K.; Harkes, J.A.; Gilbert, B.; Winger, D.G.; Golubets, K.; Akilov, O.; Satyanarayanan, M. Computer-aided classification of melanocytic lesions using dermoscopic images. *J. Am. Acad. Dermatol.* **2015**, *73*, 769–776. [CrossRef]
9. Kasmi, R.; Mokrani, K. Classification of malignant melanoma and benign skin lesions: Implementation of automatic ABCD rule. *IET Image Process.* **2016**, *10*, 448–455. [CrossRef]
10. Celebi, M.E.; Kingravi, H.A.; Uddin, B.; Iyatomi, H.; Aslandogan, Y.A.; Stoecker, W.V.; Moss, R.H. A methodological approach to the classification of dermoscopy images. *Comput. Med. Imaging Graph.* **2007**, *31*, 362–373. [CrossRef]
11. Abuzaghleh, O.; Barkana, B.D.; Faezipour, M. Noninvasive Real-Time Automated Skin Lesion Analysis System for Melanoma Early Detection and Prevention. *IEEE J. Transl. Eng. Health Med.* **2015**, *3*, 4300212. [CrossRef]
12. Bozorgtabar, B.; Sedai, S.; Roy, P.K.; Garnavi, R. Skin lesion segmentation using deep convolution networks guided by local unsupervised learning. *IBM J. Res. Dev.* **2017**, *61*, 6:1–6:8. [CrossRef]
13. Premaladha, J.; Ravichandran, K.S. Novel Approaches for Diagnosing Melanoma Skin Lesions Through Supervised and Deep Learning Algorithms. *J. Med. Syst.* **2016**, *40*, 96. [CrossRef]
14. Celebi, M.E.; Iyatomi, H.; Stoecker, W.V.; Moss, R.H.; Rabinovitz, H.S.; Argenziano, G.; Soyer, H.P. Automatic detection of blue-white veil and related structures in dermoscopy images. *Comput. Med. Imaging Graph.* **2008**, *32*, 670–677. [CrossRef]
15. Barata, C.; Marques, J.S.; Rozeira, J. A System for the Detection of Pigment Network in Dermoscopy Images Using Directional Filters. *IEEE Trans. Biomed. Eng.* **2012**, *59*, 2744–2754. [CrossRef]
16. Abbas, Q.; Celebi, M.E.; García, I.F. Hair removal methods: A comparative study for dermoscopy images. *Biomed. Signal Process. Control* **2011**, *6*, 395–404. [CrossRef]
17. Toossi, M.T.B.; Pourreza, H.R.; Zare, H.; Sigari, M.-H.; Layegh, P.; Azimi, A. An effective hair removal algorithm for dermoscopy images. *Ski. Res. Technol.* **2013**, *19*, 230–235. [CrossRef]

18. Liu, Z.-Q.; Cai, J.-H.; Buse, R. *Hand-writing Recognition: Soft Computing and Probablistic Approaches*, 1st ed.; Springer: Berlin/Heidelberg, Germany, 2003.
19. Rakowska, A. Trichoscopy (hair and scalp videodermoscopy) in the healthy female. Method standardization and norms for measurable parameters. *J. Dermatol. Case Rep.* **2019**, *3*, 14. [CrossRef]
20. Ma, Z.; Tavares, J.M.R.S. A Novel Approach to Segment Skin Lesions in Dermoscopic Images Based on a Deformable Model. *IEEE J. Biomed. Health Inform.* **2016**, *20*, 615–623. [CrossRef]
21. Weatherall, I.L.; Coombs, B.D. Skin Color Measurements in Terms of CIELAB Color Space Values. *J. Investig. Dermatol.* **1992**, *99*, 468–473. [CrossRef]
22. Chan, T.F.; Vese, L.A. Active contours without edges. *IEEE Trans. Image Process.* **2001**, *10*, 266–277. [CrossRef]
23. Tamura, H.; Mori, S.; Yamawaki, T. Textural Features Corresponding to Visual Perception. *IEEE Trans. Syst. Man Cybern.* **1978**, *8*, 460–473. [CrossRef]
24. Lee, T.K.; McLean, D.I.; Atkins, M.S. Irregularity index: A new border irregularity measure for cutaneous melanocytic lesions. *Med. Image Anal.* **2002**, *7*, 47–64. [CrossRef]
25. Eltayef, K.; Li, Y.; Liu, X. Detection of Pigment Networks in Dermoscopy Images. *J. Physics Conf. Ser.* **2017**, *787*, 012033. [CrossRef]
26. Platt, J. Probabilistic outputs for support vector machines and comparisons to regularized likelihood methods. In *Advances in Large Margin Classifiers*; The MIT Press: Cambridge, MA, USA, 2000; pp. 61–74.
27. Pathan, S.; Prabhu, K.G.; Siddalingaswamy, P.C. Hair detection and lesion segmentation in dermoscopic images using domain knowledge. *Med. Biol. Eng. Comput.* **2018**, *56*, 2051–2065. [CrossRef]
28. Mendonca, T.; Ferreira, P.M.; Marques, J.S.; Marcal, A.R.S.; Rozeira, J. PH^2-A dermoscopic image database for research and benchmarking. In Proceedings of the 2013 35th Annual International Conference of the IEEE Engineering in Medicine and Biology Society (EMBC), Osaka, Japan, 3–7 July 2013; pp. 5437–5440. [CrossRef]
29. ISIC 2016: Skin Lesion Analysis Towards Melanoma Detec-tion. Available online: https://challenge.kitware.com/#challenge/n/ISBI_2016%3A_Skin_Lesion_Analysis_Towards_Melanoma_Detection (accessed on 24 September 2017).
30. Codella, N.C.F.; Gutman, D.; Celebi, M.E.; Helba, B.; Marchetti, M.A.; Dusza, S.W.; Kalloo, A.; Liopyris, K.; Mishra, N.; Kittler, H.; et al. Skin Lesion Analysis Toward Melanoma Detection: A Challenge at the 2017 International Symposium on Biomedical Imaging (ISBI), Hosted by the International Skin Imaging Collaboration (ISIC). *arXiv* **2017**, arXiv:1710.05006.
31. Yu, L.; Chen, H.; Dou, Q.; Qin, J.; Heng, P.-A. Automated Melanoma Recognition in Dermoscopy Images via Very Deep Residual Net-works. *IEEE Trans. Med. Imaging* **2017**, *36*, 994–1004. [CrossRef]
32. Pathan, S.; Prabhu, K.G.; Siddalingaswamy, P. A methodological approach to classify typical and atypical pigment network patterns for melanoma diagnosis. *Biomed. Signal Process. Control* **2018**, *44*, 25–37. [CrossRef]
33. Pennisi, A.; Bloisi, D.D.; Nardi, D.; Giampetruzzi, A.R.; Mondino, C.; Facchiano, A. Skin lesion image segmentation using Delaunay Triangulation for melanoma detection. *Comput. Med. Imaging Graph.* **2016**, *52*, 89–103. [CrossRef]

applied sciences

MDPI

Article

Automatic Breast Tumor Screening of Mammographic Images with Optimal Convolutional Neural Network

Pi-Yun Chen [1], Xuan-Hao Zhang [1], Jian-Xing Wu [1], Ching-Chou Pai [1,2], Jin-Chyr Hsu [3], Chia-Hung Lin [1,*] and Neng-Sheng Pai [1,*]

1 Department of Electrical Engineering, National Chin-Yi University of Technology, Taichung City 41170, Taiwan; chenby@ncut.edu.tw (P.-Y.C.); jefi889c1ckqw2012@gmail.com (X.-H.Z.); jian0218@gmail.com (J.-X.W.); actirin1945@gmail.com (C.-C.P.)
2 Division of Cardiovascular Surgery, Show-Chwan Memorial Hospital, Changhua 500, Taiwan
3 The Consultant Physician of Taichung Hospital of the Ministry of Health and Welfare, Taichung City 403, Taiwan; jinchyr.hsu@msa.hinet.net
* Correspondence: eechl53@gmail.com (C.-H.L.); pai@ncut.edu.tw (N.-S.P.)

Abstract: Mammography is a first-line imaging examination approach used for early breast tumor screening. Computational techniques based on deep-learning methods, such as convolutional neural network (CNN), are routinely used as classifiers for rapid automatic breast tumor screening in mammography examination. Classifying multiple feature maps on two-dimensional (2D) digital images, a multilayer CNN has multiple convolutional-pooling layers and fully connected networks, which can increase the screening accuracy and reduce the error rate. However, this multilayer architecture presents some limitations, such as high computational complexity, large-scale training dataset requirements, and poor suitability for real-time clinical applications. Hence, this study designs an optimal multilayer architecture for a CNN-based classifier for automatic breast tumor screening, consisting of three convolutional layers, two pooling layers, a flattening layer, and a classification layer. In the first convolutional layer, the proposed classifier performs the fractional-order convolutional process to enhance the image and remove unwanted noise for obtaining the desired object's edges; in the second and third convolutional-pooling layers, two kernel convolutional and pooling operations are used to ensure the continuous enhancement and sharpening of the feature patterns for further extracting of the desired features at different scales and different levels. Moreover, there is a reduction of the dimensions of the feature patterns. In the classification layer, a multilayer network with an adaptive moment estimation algorithm is used to refine a classifier's network parameters for mammography classification by separating tumor-free feature patterns from tumor feature patterns. Images can be selected from a curated breast imaging subset of a digital database for screening mammography (CBIS-DDSM), and K-fold cross-validations are performed. The experimental results indicate promising performance for automatic breast tumor screening in terms of recall (%), precision (%), accuracy (%), F1 score, and Youden's index.

Keywords: convolutional neural network (CNN); fractional-order cconvolutional operation; adaptive moment estimation algorithm

Citation: Chen, P.-Y.; Zhang, X.-H.; Wu, J.-X.; Pai, C.-C.; Hsu, J.-C.; Lin, C.-H.; Pai, N.-S. Automatic Breast Tumor Screening of Mammographic Images with Optimal Convolutional Neural Network. *Appl. Sci.* **2022**, *12*, 4079. https://doi.org/10.3390/app12084079

Academic Editors: Michalis Vrigkas, Christophoros Nikou and Ioannis A. Kakadiaris

Received: 13 February 2022
Accepted: 15 April 2022
Published: 18 April 2022

1. Introduction

As per statistics provided in 2020 by Taiwan's Ministry of Health and Welfare, cancer (malignant tumors) is the primary cause of death among Taiwanese people. In recent years, breast cancer (BC) in females is among the top four cancers (first place) and is one of the diseases that most definitely cannot be ignored. The age at which women possibly develop BC is between 45 and 69 years. As per latest figures on the cause of death from the Ministry of Health and Welfare and cancer registration data from the National Health Agency [1], the standardized incidence and mortality rates of female BC are 69.1 and 12.0 (per 100,000 people), respectively. Each year, more than

10,000 women are diagnosed with BC and more than 2000 women die from it. This is about 31 women being diagnosed with BC every day and six women losing their precious lives because of BC. BC is a malignant tumor that grows from epithelial cells or lobules of the breast, thus resulting in excessive tissue accumulation or hyperplasia. Ductal carcinoma and lobular carcinoma are common types of BC. When proliferating cells mutate or lose control, they can invade or destroy other adjacent tissues and organs, transfer to other organs via the blood or the lymphatic system, and cause breast pain. Symptoms include irregular lumps (partial) on breasts, sunken skin, orange peel skin, redness, ulceration, abnormal secretions, venous vasodilation, and swollen lymph nodes under the arms. For clinical testing, visual inspection or palpation helps identify the existence and location of the tumor (hard mass). However, it is not immediately clear whether the tumor is benign or malignant or whether metastasis is happening. Either additional diagnostic tools or instruments must be used for verification. As per statistics, there is improved chance of tumors on the right side of the breast. The most common location is the upper right corner of the breast, accounting for about 26% of all occurrences. As per the size of the tumor and severity of lymph node (sentinel lymph node) metastasis, BC can be divided into four stages. If the abnormality is quickly detected, the survival rate is higher, and the treatment is more effective. Hence, the early symptoms of BC are traceable, whereas the survival rate of BC, if detected early, can be larger than 90%.

Accordingly, in recent years, artificial intelligence (AI) methods and the collection of data for big data (BD) are playing increasingly important roles for automatic tumor screening, such as liver, lung, and breast cancers. AI methods include models based on machine learning, deep learning, and broad learning [2–9]. For example, in Taiwan, data collection for BD includes medical images of relevant major diseases that major medical centers have been actively collecting both locally and abroad in recent years (since 2018). These medical images comprise fifteen categories, including X-ray images, angiography, magnetic resonance imaging (MRI), and computer tomography (CT). Additionally, a database such as CBIS-DDSM (Curated Breast Imaging Subset of Digital Database for Screening Mammography), is a database containing approximately 2500 enrolled subjects, employed for the studies of the mammographic classification of breast lesions. This classification consists of normal cases, benign cases, malignant cases, and pathology information [10], which provides ground truth validation that makes the DDSM applicable in the development and validation of the decision support systems. The region of interest (ROI) and pathological information can be gathered continuously in the current database. Coupled with the improvement of information communication, digital data processing, AI methods, and machine vision models, software and hardware equipment can process the large amount of digital data [11–13]. However, a decision support system with computer-aided diagnosis and detection algorithms for breast tumor screening in mammography are needed. Currently, regarding AI algorithms, the multilayer machine vision recognition model that is composed of convolutional neural networks (CNNs) is the most commonly used model. It can be used for digital image processing, such as feature enhancement, feature extraction, data simplification, and pattern recognition task [2–9,13–15]. Hence, this study uses breast mammogram images to establish a two-dimensional (2D) fractional order-based CNN classifier that can automatically perform breast tumor screening tasks in clinical applications. We expect that a set of automatic screening tools suitable for clinical usage that takes in mammogram images can be developed to achieve the rapid identification of whether a tumor is malignant or benign. When any breast tumor is suspected, the results of the rapid screening can be used as a basis of reference for subsequent fine needle aspiration cytology/biopsy examinations. Hence, this assistive tool solves the problem of insufficient human resources for manual screening, which can potentially lead to additional problems in the medical diagnosis process. Solving this procedural congestion allows clinicians to focus more on follow-up medical strategies.

The 2D CNNs may comprise several convolutional-pooling layers and a fully connected network in the classification layer, such as back-propagation neural networks and Bayesian networks, which combine the image enhancement, feature extraction, and classification tasks to form an individual scheme [16,17] which achieves promising accuracy for image classification in breast tumor screening. These CNNs are usually greater than 10 convolutional-pooling layers which perform the above mentioned image preprocessing and postprocessing tasks and then increase the identification accuracy. Hence, this multilayer design may gradually replace machine-learning (ML) methods [18,19], which perform image segmentation and feature extraction as an image preprocess for mammograms and breast MRIs and then use the fixed features obtained to train a classifier. Both CNN and ML-based image segmentation [20] can learn the specific features or knowledge representations to automatically identify the boundaries of ROI and then detect the breast lesions. Traditional ML methods have fewer parameters that can easily be optimized by the gradient descent optimization or back-propagation algorithms through training with small-to-medium-sized datasets [21,22]. Through a series of convolutional and pooling processes, the multilayer CNN can enhance and extract the desired object at different scales and different levels from low-level features (extract object's edge) to high-level information (extract object's shape) for detecting nonlinear features, which can increase nonlinearity and obtain feature representation. Then, the pooling process with maximum pooling (MP) is used to reduce the sizes of feature maps for obtaining abstract features. Thus, in contrast to the traditional machine-learning method, CNN-based methods can learn to extract the feature patterns from the raw data and improve the classification accuracy significantly. However, small- or medium-sized datasets are insufficient to train a deep-learning-based CNN. For example, from the existing literature [23–26], such as AlexNet (eight-layer CNN) [25] and ZFnet [26], it can be observed that the deep-learning-based CNN requires several convolutional-pooling layers and fully connected layers for the large-scale image classification (ImageNet image database [27,28]). This CNN can learn to optimize features during the training stage, process large inputs with sparsely connected weights, adapt to different sizes of 2D images, and reduce error rates. Furthermore, this approach demonstrates greater computational efficiency compared with the traditional fully connected multilayer perceptron (MLP) networks. Despite its many advantages, however, a deep-learning-based CNN presents several drawbacks and limitations, such as the number of convolutional-pooling layers' determination, the number of convolutional windows and pooling windows, the sizes of convolutional window assignment (3×3, $5 \times 5, 7 \times 7, 9 \times 9$, 11×11), the high computational complexity and large-scale dataset requirement for training the CNN-based classifier, and the poor suitability for real-time applications. Additionally, multi-convolutional-pooling processes with different sizes of convolution masks will result in a very large information loss for feature extraction, and this will result in increased complexity levels. The multilayer CNN must be performed with a graphics processing unit (GPU) to speed up the training and classification tasks by making use of a large amount of training and testing data.

Therefore, to simplify the image processing and classification tasks, this study aimed to design a suitable number of convolutional-pooling layers and a classification layer that is capable of increasing the identification accuracy of image classification, to facilitate automatic breast tumor screening. As observed from Figure 1, we utilized a multilayer classifier, consisting of a fractional-order convolutional layer, two convolutional-pooling layers, a flattening layer, and a multilayer classifier in the classification layer. In the first convolutional layer, a 2D spatial convolutional process with two 3×3 fractional-order convolutional masks was used to perform the enhancement task and to remove unwanted noise from the original mammography image, to distinguish the edges and shapes of the object. In the second and third convolutional layers, sixteen 3×3 kernel convolutional windows were used to subsequently enhance and sharpen the feature patterns twice; hence, the tumor contour could easily be highlighted and distinguished for feature pattern extraction. Consequently, two MP processes were used to reduce the dimensions of the feature patterns, which conducted network training to avoid failing in overfitting problems [29,30]. In the classification layer, a multilayer classifier with an input layer, two hidden layers, and an output layer is implemented to perform the pattern recognition task, which separates tumor-free feature patterns from tumor feature patterns. To reduce the error rates, an adaptive moment estimation method (ADAM) can compute the adaptive learning rates for updating network parameters by storing an exponentially decaying average of past squared gradients [31,32], which combines two stochastic gradient descent approaches, including adaptive gradients and root mean square propagation. Its optimization algorithm uses randomly selected training data subsets to compute the gradient, instead of using the entire dataset. The momentum term can speed up the gradient descent by converging faster. The ADAM algorithm has a simple implementation, computation efficiency, and fewer memory requirements, and is appropriate for operations with large datasets and parameters for training the multilayer CNN models. A total of 78 subjects is selected from the MIAS (Mammographic Image Analysis Society) Digital Mammogram Database (United Kingdom National Breast Screening Program) for experimental analysis. The clinical information was confirmed and agreed upon by expert radiologists for biomarkers, such as image size, image category, background tissue, class of abnormality, and severity of abnormality [33,34]; the image database included a total of 156 mammography images (including right and left images), including 94 normality cases and 62 abnormalities involving benign and malignant cases. The ROIs were extracted by a 100 × 100 bounding box, and then the 932 feature patterns were extracted by using the proposed convolutional-pooling processes including 564 abnormalities and 368 tumor-free patterns. By making use of cross-validation, the dataset was randomly divided into two halves: 50% of the dataset was used for training the classifier, and 50% of the dataset was used for evaluating the classifier's performance. Thus, tenfold cross-validation is used to verify the performances of the proposed multilayer deep-learning-based CNN with the proposed convolutional-pooling layers in terms of recall (%), precision (%), accuracy (%), F1 score, and Youden's index [35,36]. Therefore, the optimal architecture of multilayer CNN can be determined, and may potentially be applied to establish a classifier for automatic breast tumor screening in clinical applications.

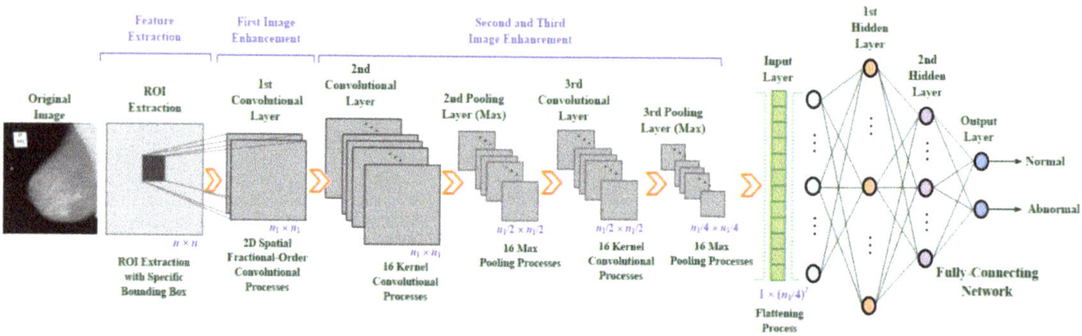

Figure 1. Architecture of the proposed multilayer CNN for automatic breast tumor screening.

The remainder of this study is organized as follows: Section 2 describes the methodology, including the design of the multilayer deep-learning-based CNN, the adaptive moment estimation method, the classifier's performance evaluations, and the computer assistive system. Sections 3 and 4 present the experimental setup, testing of different multilayer CNN models and determination of the suitable CNN architecture, testing of the first convolutional layer for image enhancement, determination of the mask types, feasibility tests, and experiment results for clinical applications, and the conclusions, respectively.

2. Materials and Methods

2.1. Design of the Multilayer Deep-Learning-Based CNN

Multilayer deep-learning-based CNN includes the multiple convolutional layers, pooling layers, and pattern recognition layers (classification layers). It is a multilayer model for image classification that combines multiple functions, such as image feature enhancement, feature extraction, parameter simplification, and pattern recognition. In recent years, deep-learning and broad-learning technologies have been applied for medical image processing, segmentation, and classification [3–9,13–15]. Compared with traditional MLP networks, CNN can process feature enhancement and extraction at the front end of the network; therein lies its advantage in processing 2D medical images. However, image processing generates a considerable number of parameters, increasing the amount of computations and the time required for computing. With pooling processes, the number of features is reduced in the middle of the network to reduce the computational time. Finally, the image classification layer assists with screening tasks using a pattern to separate the normality from abnormalities. This study uses the multilayer CNN model to develop an automatic breast tumor screening based on 2D mammogram images, as shown in Figure 1. This automatic screening process includes: (1) Region of interest (ROI) extraction; (2) feature enhancement and extraction; and (3) rapid screening of breast tumors. Each function is described as follows:

ROI extraction: In this study, we use digital data in the mammogram X-ray image database (161 female patients, 322 images) [33,34] provided by the MIAS. The biomarkers of MIAS database clearly mark the positions and tumor sizes [33]. The statistical results of probability distribution of tumor locations, in accordance with the MIAS database's biomarkers, are shown in Figure 2a. Looking at the distribution probability of tumor locations, the most frequent location of tumors is the right and the upper outer quadrant of the breast. Given a 2D image with 4320 pixels $\times$ 2600 pixels, we define the priority for automatically extracting ROI with a specific bounding box based on the distribution probability. Areas with greater probability are first in line for ROI extraction, and the priority order is stored in the work queue. The priority order is shown in Figure 2b. ROI image extraction and tumor detection is performed as per the priority order.

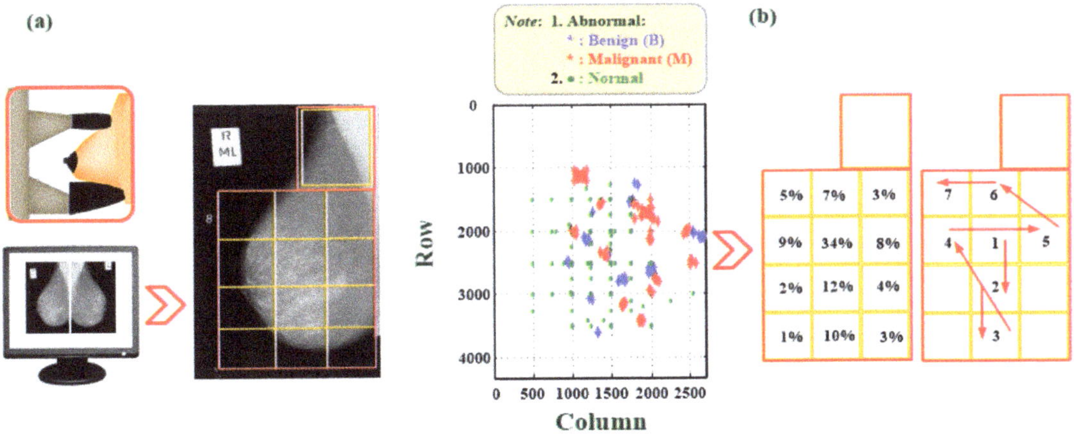

Figure 2. ROI block cutting and priority extraction. (**a**) The statistics of the prevalence of malignant and benign tumors; (**b**) priority of ROI block for automatic ROI extraction.

- Feature enhancement and extraction: A multilayer 2D convolution operation is used to magnify the texture of what might be tumor tissue and edge information (usually two or more layers are used), as shown in Figure 2a. Each layer uses a 3×3 sliding window to perform the operation of the convolutional weight. First, a 2D fractional convolution operation is performed to magnify the tumor characteristics. Then, by combining multilayer convolutional weight calculations, the contour of the tumor is gradually strengthened, noise is removed, and the image is sharpened. These effects help strengthen the target area and retain non-characteristic information. This study applies the 2D spatial fractional-order convolutional processes in the fractional convolutional layer, selects the appropriate fractional order parameters, and performs convolution in the x and y directions, thus yielding a combination of 2D weight values in space, the general formula being [35–38]:

$$C^{v}I_{xy} = conv(I_{xy}, M(i,j), v)^{T} \tag{1}$$

$$C_{x}^{v}I_{xy} = \sum_{i=-\frac{h-1}{2}}^{\frac{h-1}{2}} \sum_{j=-\frac{h-1}{2}}^{\frac{h-1}{2}} M_{x}(i,j)I(x+i,y+j) \tag{2}$$

$$C_{y}^{v}I_{xy} = \sum_{j=-\frac{h-1}{2}}^{\frac{h-1}{2}} \sum_{i=-\frac{h-1}{2}}^{\frac{h-1}{2}} M_{y}(j,i)I(x+j,y+i) \tag{3}$$

where $h = 3$ is the dimension of the convolution window, v is a fractional parameter and $v \in (0, 1)$, and $I(x, y) \in [0, 255]$ is the pixel value at point (x, y) in a 2D image. Each fractional-order convolutional mask multiplies each element, $M(i, j)$ or $M(j, i)$, by the corresponding input pixel values, $I(x, y)$, and then obtains an enhanced feature pattern containing spatial features in the x-axis and y-axis directions. These 2D spatial convolutional processes act as two low-pass frequency filters [39] and then remove the high-spatial-frequency components from a breast mammogram. In this study, the image dimension is $n \times n$, $x = 1, 2, 3, \ldots, n$, and $y = 1, 2, 3, \ldots, n$. M_x and M_y are 3×3 convolutional windows that can be written as follows [35–38]:

$$M_{x} = \begin{bmatrix} 0 & \frac{v^2-v}{2} & 0 \\ 0 & -v & 0 \\ 0 & 1 & 0 \end{bmatrix}, \; M_{y} = M_{x}^{T} = \begin{bmatrix} 0 & 0 & 0 \\ \frac{v^2-v}{2} & -v & 1 \\ 0 & 0 & 0 \end{bmatrix} \tag{4}$$

where $v \in (0, 1)$ is the fractional-order parameter. A sliding stride = 1 is selected for spatial domain-based convolution operations in the horizontal and vertical directions. The results of the convolution operation of (1) and (2) are combined and normalized, and the approximate formula is written below:

$$\nabla^{v}I_{xy} = \begin{bmatrix} C_{x}^{v}I_{xy} \\ C_{y}^{v}I_{xy} \end{bmatrix}^{T}, \; \left|\nabla^{v}I_{xy}\right| \cong \frac{\left|C_{x}^{v}I_{xy}\right| + \left|C_{y}^{v}I_{xy}\right|}{255} \tag{5}$$

These multilayer convolutional layers are also called the perception layers of the CNN network for feature enhancement and extraction. After feature extraction, the 2×2 sliding window is used to perform maximum pooling (MP), as shown in the general formula (6):

$$MP\Big|_{\frac{n}{2}\times\frac{n}{2}} = \max_{M_{2\times 2}}\left(\left|\nabla^{v}I_{xy}\right|\Big|_{n\times n}\right) \tag{6}$$

After MP, the number of feature patterns is reduced to 25% of the total number of original feature images. This reduction in the dimensions of the feature patterns can overcome the overfitting problem for training a multilayer classifier.

- Rapid screening of breast tumors: Breast tumors can be identified at the image classification layer, which includes the flattening process (FP) and a multilayer classifier, as seen in Figure 1. The FP can convert a 2D feature matrix into a 1D feature vector, which is then fed as the input vector of the classifier for further pattern recognition.

After two MP treatments, the FP treatment may be written as shown in the general formula (7):

$$X\Big|_{1\times(\frac{n}{4})^2} = FP(MP\Big|_{\frac{n}{4}\times\frac{n}{4}}) \tag{7}$$

where X is the 1D feature vector of the multilayer classifier used as input. In this study, the multilayer classifier includes an input layer, two hidden layers (i.e., the first and second hidden layers), and an output layer.

In two hidden layers, the Gaussian error linear unit (GeLU) function [40–42] is used as the activation function in each hidden node. This activation function performs a nonlinear conversion, as shown in Figure 3, which can be expressed as follows:

$$GeLU(x_i) = 0.5x_i(1 + \tanh(\sqrt{\frac{2}{\pi}}(x_i + 0.4472{x_i}^3))),\ i\ = 1,\ 2,\ 3 \tag{8}$$

where x_i is the 1D feature vector used as input, $i = 1, 2, 3, \ldots, n$, $X = [x_1, x_2, x_3, \ldots, x_n]$. The training of the multilayer classifier uses the back-propagation algorithm to adjust the connecting weighted parameters of the classifier and set the loss function as the convergent condition for terminating the training stage. For multi-class classification, multiple classes of binary cross-entropy functions [7,43–45] are shown in Equation (9):

$$L = -\frac{1}{K}\sum_{j=1}^{m}\sum_{k=1}^{K} t_{j,k}\log_2(y_{j,k}) + (1 - t_{j,k})\log_2(1 - y_{j,k}),\ j\ = 1,\ 2,\ 3 \tag{9}$$

$$Y = GeLU(XW) \tag{10}$$

where $t_{j,k}$ is the target value (desired class), $T = [t_{1,k}, t_{2,k}, t_{3,k}, \ldots, t_{m,k}]$ for multiple classes; $y_{j,k}$ is the outputted prediction value, $Y = [y_{1,k}, y_{2,k}, y_{3,k}, \ldots, y_{m,k}]$; and m is the number of classifications. This study sets $m = 2$, either normal or abnormal, coding as $Y = [1, 0]$ and $Y = [0, 1]$, respectively, $k = 1, 2, 3, \ldots, K$, is the number of training data, and W is the weighted parameter matrix of the classifier with a fully connecting network.

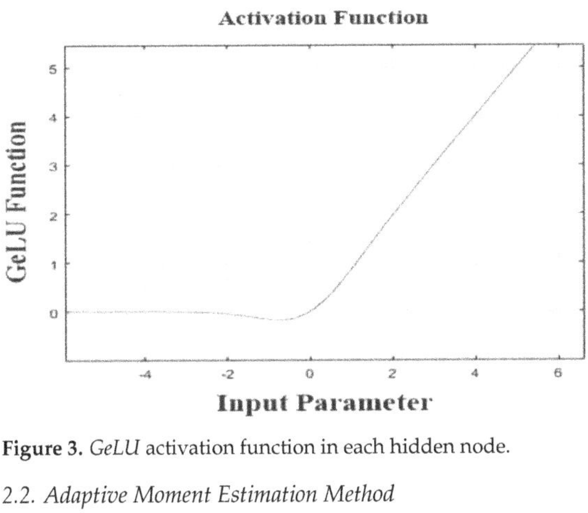

Figure 3. *GeLU* activation function in each hidden node.

2.2. Adaptive Moment Estimation Method

In the classification layer, the network connecting weighted parameters are adjusted by using BPA to minimize the loss function. The smaller the cross-entropy value, the smaller the classification error rate, and the higher the accuracy that can be obtained. The adjustment formula of weight parameters of the classifier uses the adaptive moment estimation (ADAM) optimization method, as follows [31,46]:

$$w(p+1) = w(p) + \eta\frac{\hat{m}(p)}{\sqrt{\hat{v}(p)} + \delta} \tag{11}$$

where $\hat{m}(p) = \frac{m(p)}{1-\beta_1}$ and $\hat{v}(p) = \frac{v(p)}{1-\beta_2}$ are adjustment parameters; η is the learning rate of the classifier; δ is the smoothing value; β_1 = 0.900 and β_2 = 0.999 are the attenuation rates of each iteration; p = 1, 2, 3, ... , p_{max}; and p_{max} is the maximum number of iterations. Each iteration computation adjusts the weighted parameters of the classifier within a limited range with the parameters of Equation (11), as shown in Equations (12) and (13) [31,46]:

$$m(p) = \beta_1 m(p-1) + (1-\beta_2)\frac{\partial L}{\partial w} \quad (12)$$

$$v(p) = \beta_1 v(p-1) + (1-\beta_2)(\frac{\partial L}{\partial w})^2 \quad (13)$$

With the above-mentioned formulas, the best parameters can be quickly obtained using matrix operations and the loss function (9) can be minimized.

The proposed multilayer classifier used in this study is a fully connecting network. The number of nodes in the hidden layer in the middle of the network is set as per the graph's data type and complexity. Optimization algorithms are used to adjust the connecting weighted parameters of the classifier to minimize the loss function. The input ROI image size used in this study is 100 pixels × 100 pixels. There is one fractional-order convolutional layer, two convolutional layers, two maximum pooling layers, a flattening layer, and a fully connected multilayer classifier. The relevant information about the proposed multilayer CNN is shown in Table 1.

Table 1. Relevant information about the proposed multilayer CNN.

Layer Function	Manner	Feature Pattern
Input Feature Pattern	ROI Extraction with 100 × 100 Bounding Box	2D, 100 pixels × 100 pixels
1st Convolutional Layer	3 × 3 Fractional-Order Convolutional Window, Stride = 1	2D, 100 pixels × 100 pixels
2nd Convolutional Layer	3 × 3 Kernel Convolution Window, Stride = 1	2D, 100 pixels × 100 pixels
2nd Pooling Layer	2 × 2 Maximum Pooling Layer, Stride = 2	2D, 50 pixels × 50 pixels
3rd Convolutional Layer	3 × 3 Kernel Convolution Window, Stride = 1	2D, 50 pixels × 50 pixels
3rd Pooling Layer	2 × 2 Maximum Pooling Layer, Stride = 2	2D, 25 pixels × 25 pixels
Flattening Layer	Flattening Process	1D, 1 × 625 feature vector
Classification Layer	Multilayer Classifier: 625 input nodes, 168 hidden nodes 64 hidden nodes, 2 output nodes (for normality and abnormality) Algorithm: ADAM Algorithm	1 × 625 Feature Vector Feeding into Multi-Layer Classifier

2.3. Classifier's Performance Evaluations

This study uses the cross-validation ten times to evaluate the performance of the proposed multilayer CNN-based classifier. Each time, the dataset is divided into the two groups of normal and abnormal feature patterns. The dataset is then randomly divided into two halves: 50% training dataset and 50% testing dataset. Repeat the procedure ten folds to confirm the performances of the proposed classifier, such as the evaluation indicators shown in the formulas for precision (%), recall (%), F1 score, and accuracy (%) [35,36,47] in Table 2. For each fold cross-validation, the multilayer CNN-based classifier will produce a confusion matrix comprising four parameters, including TP (True Positive), FP (False Positive), TN (True Negative), and FN (False Negative). These parameters help to determine the indexes for evaluating the performances of the proposed classifier. The precision (%) indicator represents the rate that a TP can be correctly identified among all TPs (positive samples). Generally, a model must be larger than 80% to be recognized as a good classifier. The recall (%) indicator is defined as TP/(TP + FN). The F1 Score ($\in$ [0,1]) is the harmonic mean of precision (%) and recall (%) indexes. Its index combines the two in a single evaluation index. The higher the value of the F1 score and the closer it is to 1, the better the classifier is at prediction.

Table 2. Formulas for the evaluation criteria of the proposed classifier, including precision (%), recall (%), accuracy (%), and F1 score.

	Actual		Total	Precision (%)
Predicted	TP	FP	TP + FP	(TP)/(TP + FP)
	FN	TN	FN + TN	
Total	TP + FN	FP + TN	Accuracy (%): (TP + TN)/(TP + FP + TN + FN)	
Recall (%)	(TP)/(TP + FN)			
F1 Score	(2TP)/(2TP + FP + FN)			

2.4. Computer Assistive System for Automatic Breast Tumor Screening

This study uses the LabVIEW 2019 (NI^{TM}) software to develop a computer assistive system for automatic breast tumor screening, integrating: (1) ROI image extraction, (2) feature enhancement and extraction, and (3) breast tumor screening classifier and other functions. Algorithms for functions (1) and (2) are developed using the MATLAB Script tool. The multilayer CNN algorithm and the interface shown in Figure 4 are written by Python software. The interface works as follows:

- Zone ①: Sets the source path of breast mammography images;
- Zone ②: Loads and displays the selected mammography images;
- Zone ③: As per the priority order, extract ROI images and perform automatic tumor screening. In this study, six areas at which tumors are most possibly identified are designated. The CAS automatically prioritizes the ROI cutting feature patterns (100 pixels × 100 pixels), as seen in Figure 2, and then screens those areas. The block marked ③ can show the output of the classifier, the identification result, and the classification information. The red and green circles show the normality and abnormality. The output value of the classifier must be >0.5 to have a high degree of confidence that there is a suspected breast tumor.

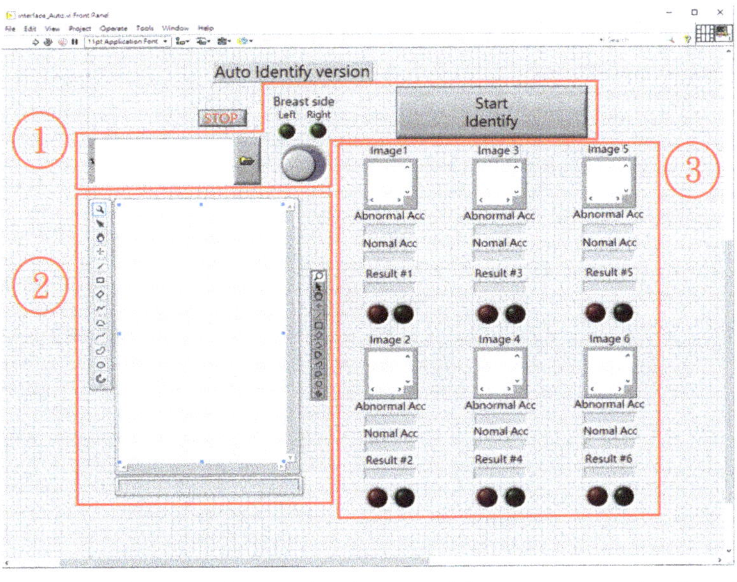

Figure 4. The human machine interface for automatic breast tumor screening: Zone ①: Sets the source path of images, Zone ②: Loads and displays the images, Zone ③: Extract ROI images and perform automatic tumor screening.

The human machine interface designed in this study can be used in clinical applications when switched to manual mode. The user can manually select six ROI blocks and screenshots and then save these images in a temporary database. When the number of screenshots reaches the set number (i.e., default of six ROIs), the multilayer CNN classifier performs the classification task as per the priority order of the queue and returns the identification results. The clinician then receives messages to confirm the existence of a possible tumor.

2.5. Experimental Setup

In the MIAS database, the most common image size was 4320 pixels × 2600 pixels. Thus, in this study, the dimensions 4320 pixels × 2600 pixels were selected for breast tumor screening [33,34]. The vertical and horizontal resolutions of each image were identical at 600 dpi, with a bit depth of 24 bits. A total of 156 mammography images (78 subjects), including 62 images with malignant (M) or benign (B) tumors and 94 images without tumors, were obtained. Given a specific bounding box measuring 100 pixels × 100 pixels, feature patterns were screenshots from the 156 images. In total, 932 feature patterns, including 564 tumors and 368 tumor-free screenshots, were obtained. In each classifier's training stage, 282 tumor and 184 tumor-free screenshots (50% feature patterns) were randomly selected to train the multilayer CNN classifier. The remaining 50% of the feature patterns were used to evaluate the classifier's performance for each cross-validation. This study used the relevant data, as shown in Table 1, to establish a multilayer CNN-based classifier. We designed a fractional-order convolutional layer, two general convolutional layers, and two MP processing layers for feature enhancement and extraction. The convolutional layer had 16 kernel convolutional windows. In the kernel window, the sliding window moved the number of columns and rows in steps of 1 (stride = 1) at each point of the convolution operation. The padding parameter was set to 1 to maintain the feature pattern after the convolutional operation. During the pooling process, the MP window moved with a stride of 2 (stride = 2) each time. During each feature enhancement, and extraction process, the possible tumor features and contours were gradually enhanced by the convolutional-pooling processes; hence, it can be observed that the multilayer CNN-based classifier can improve the accuracy of pattern recognition based on these enhanced features. Tenfold cross-validation was performed using precision (%), recall (%), F1 score, and accuracy (%) as indicators [35,36,48] to evaluate the prediction performance of the proposed classifier. Figure 5 shows the visualization of the confusion matrix; for example, the classifier used 466 images for rapid screening, and the results show 178 TPs, 13 FPs, 269 TNs, and 6 FNs. The precision (%), recall (%), F1 score, and accuracy (%) can be calculated from the confusion matrix.

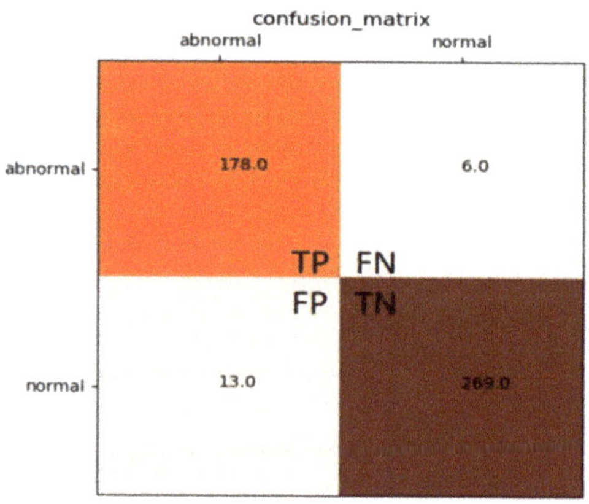

Figure 5. Output confusion matrix of the multilayer CNN-based classifier.

3. Experimental Results

This study compares the training time, accuracy, training curve, and prediction performance of multilayer CNN-based classifiers using different numbers of convolutional layers and pooling layers, different types of convolutional windows, and different sizes of convolutional window dimensions, as seen in Table 3. The comparison items are briefly described as follows:

- The number of convolutional layers and pooling layers: This study increases the number of convolutional layers and pooling layers from 1 to 5 and the sizes of convolution windows from 3 × 3 to 11 × 11. The processing windows for the pooling layers are set to 2 × 2, and the second to fifth convolutional layers have 16 kernel convolution windows to perform feature enhancement and extraction.
- The first convolution windows: This study selects three types of convolutional windows, including fractional-order ($v \in (0, 1)$), Sobel (first order, $v = 1$), and Histeq [35,36,38,47,49,50] windows to pre-enhance the feature pattern.

Table 3. Different convolutional layer models of the multilayer CNN-based classifier (Models #1–#5).

Model	1st Convolution Window	2nd Convolution Window	3rd Convolution Window	4th Convolution Window	5th Convolution Window	Stride	Padding
1	3 × 3, 2	-	-	-	-	1	1
2	3 × 3, 2	5 × 5, 16	-	-	-	1	1
3	3 × 3, 2	5 × 5, 16	7 × 7, 16	-	-	1	1
4	3 × 3, 2	5 × 5, 16	7 × 7, 16	9 × 9, 16	-	1	1
5	3 × 3, 2	5 × 5, 16	7 × 7, 16	9 × 9, 16	11 × 11, 16	1	1

In general, a multilayer CNN may have dozens of layers of convolutional layers. As shown in Table 3, this study designs five different multilayer models of convolutional layers and convolutional window sizes. It compares the training time and accuracy of five models to confirm the feasibility of the CNN model constructed in this study. Moreover, we establish three models for feature enhancement and extraction, as seen in Table 4, by combining different kernel convolutional windows and dimensions (3 × 3 and 5 × 5) and comparing the performance of these different models. These tests will help determine the best model for clinical applications in automatic breast tumor screening. In addition, we also use a multi-core personal computer (Intel® Q370, Intel® Core™ i7 8700, DDR4 2400 MHz 8G*3) as a development platform to implement the multilayer CNN-based classifier suggested in this study and use the graphics processing unit (GPU) (NVIDIA® GeForce® RTX™ 2080 Ti, 1755 MHz, 11 GB GDDR6) to speed up the time it takes for digital image processing. The feasibility study was validated as described in detail in the subsequent sections.

Table 4. Comparisons of average training CPU time and average accuracy (%) for five different CNN models.

Model	1	2	3	4	5
Training CPU Time (min)	<30	<240	<7	<10	<180
Average Accuracy (%)	90.99%	90.34%	95.92%	95.28%	95.71%

3.1. Testing of Different Multilayer CNN Models and Determination of the Most Suitable Architecture

As shown in Table 3, this study designs five models comprising 1–5 convolutional layers. The first layer is a fractional-order convolutional layer with two 3 × 3 convolutional windows; it is used to perform the 2D spatial convolution operations. The second to fifth convolutional layers use 16 kernel convolutional windows with different sizes of convolutional windows (3 × 3, 5 × 5, 7 × 7, 9 × 9, 11 × 11) [51] and 16 MP windows for feature enhancement and extraction. Finally, a fully connected network with two hidden layers using the adaptive moment estimation optimization method [46,48] adjusts the connecting weighted parameters with Equations (11)–(13) such that the predetermined classification is obtained. This study uses the same training dataset, specifically, 466 feature patterns (282 tumor and 184 tumor-free screenshots), to train and test the five different CNN models. We randomly generate initial parameters and train each model at least five times, thus recording the required training time and classification accuracy rate to compare the average training CPU time (min) and average accuracy (%) of models, as seen in Table 4. The testing results indicate that the three-layer convolutional layer model (Model #3 shown in Table 3) has an average training CPU time of lower than 7 min and an average accuracy (%) of larger than 95% with 466 untrained feature patterns. While the average accuracy (%) of the four- and five-layer models can reach larger than 95%, these two models require more training CPU time. Therefore, Model #3 is the most suitable model for developing an automatic breast tumor screening classifier in clinical applications.

3.2. Testing of the First Convolutional Layer and Determination of the Window Type

As seen in Table 5, three types of convolution windows in the first convolutional layer, including fractional-order windows, Sobel windows, and Histeq windows, are used perform the 2D spatial convolutions [35,36,47,49–52]. Figure 6a shows the original image and image enhancement results of these three types of convolution windows. Figure 6b shows the pixel grayscale value distribution map after the image is magnified. Compared with the original image grayscale value (0–255) distribution map, the convolution result of the first derivative-based Sobel convolution window [49] has a smoothing effect and is anti-noise but requires a considerable amount of calculations while performing convolutions; moreover, this window type produces a thicker edge contour, which results in lower accuracy in identifying the position of the target object. We can use a second-order-based convolutional window for feature enhancement, but this window is fairly susceptible to noise and thus unsuitable for obtaining the edge contour of the target; this type of window is generally used for binarization applications. The Histeq convolution (histogram regularization) [50] yields a histogram of the number of times each grayscale value appears. This histogram can describe the statistical information of the grayscale values of the image and allows the direct observation of the characteristics of the image, such as its brightness and contrast. It is primarily used for image segmentation and adjustment of grayscale values in the image. As shown in Figure 6b, the non-zero value of the histogram has a wide and uniform distribution, which indicates that the contrast of the image is high. The pixel value of the image may be readjusted to a value between 0 and 255 by using linear, piecewise linear, and nonlinear transformation functions [53]. These transformation functions are primarily used to magnify the contrast of the original image. The overall grayscale value distribution map shifts to the right, and the contrast of the image increases, thereby minimizing the effort required to highlight the outline of the malignant tumor, as shown in Figure 6a. However, this method is susceptible to factors such as illumination, viewing angle, and noise. The Histeq (histogram normalization) function [50] can automatically determine the grayscale transformation function and yield an output image with a uniform histogram. It is primarily used for contrast adjustments over a small range but could amplify background noise.

Table 5. Different convolutional layer models for feature enhancement and extraction (Models #1–#3).

Model	1st Convolutional Window and Window Size	2nd Convolutional Window and Window Size	3rd Convolutional Window and Window Size	Stride/ Padding	Maximum Pooling Window	Stride
1	Fractional Order, 3×3, 2	3×3or 5×5, 16	3×3 or 5×5, 16	1/1	2×2, 16	2
2	Sobel (First Order), 3×3, 2	3×3 or 5×5, 16	3×3 or 5×5, 16	1/1	2×2, 16	2
3	Histeq, 3×3, 2	3×3 or 5×5, 16	3×3 or 5×5, 16	1/1	2×2, 16	2

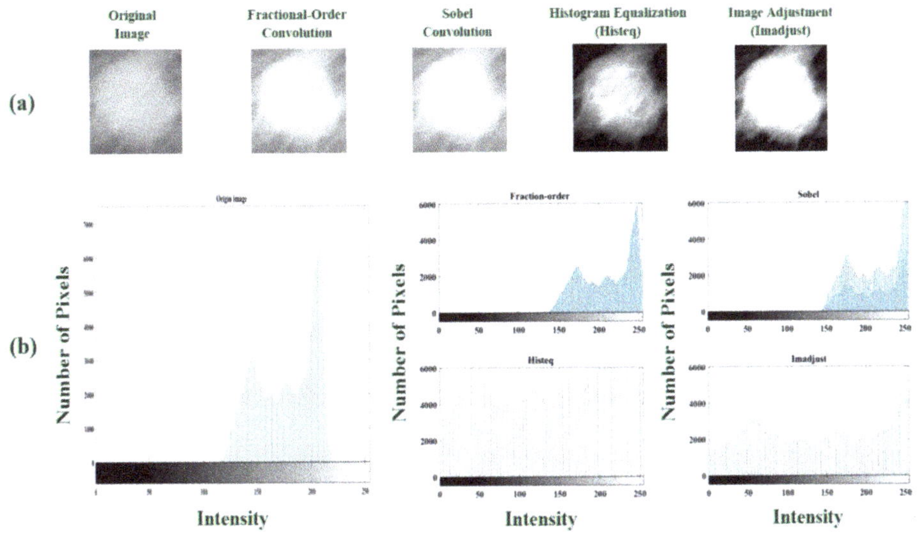

Figure 6. (**a**) The original image (malignant tumor) and the enhanced image of the three convolutions, (**b**) The original image and the pixel greyscale distribution map after image enhancement.

When fractional-order spatial convolution is conducted in 2D space, the overall grayscale value distribution moves to the right, which increases the contrast of the image and filters out noise. Thus, it shows better performance than the Sobel convolutional operation. Therefore, in the first convolutional layer, this study selects a 3 × 3 fractional-order convolution window for the first convolutional layer. In addition, the literature [35,36] proposes that setting the fractional-order parameter v = 0.30–0.40, which is also used for feature enhancement in X-ray images, could yield promising results. Thus, our study selects the parameter v = 0.35. After 2D spatial convolution and normalization operations, the contour of the sharpened target can be obtained by using Equations (1)–(5). Strengthening the target's features, retaining non-characteristic information, and removing noise are helpful for the subsequent second- and third-layer feature extraction operations and further pattern recognition.

3.3. Multilayer CNN-Based Classifier Testing and Validation

This study uses Model #1, as shown in Table 5, which adopts three convolutional layers, and the same completely connected classification layer to develop four models, as shown in Table 6. The convolutional window sizes of the second and third layers are combinations of (3 × 3, 3 × 3), (3 × 3, 5 × 5), (5 × 5, 3 × 3), and (5 × 5, 5 × 5) [51]. The image dataset is divided into two groups of equal size. The four models use 466 trained and 466 untrained feature patterns to test and confirm the performance of the classifier. A total of 1000 epochs are set for the training classifier, with the trained and untrained feature patterns. Figure 7 shows (a) the training performance of the classifier and (b) the training history curve of the classification performance validation; in (b), the solid blue line represents the results of the training performance test and the solid orange line indicates the results of classification performance validation. As the number of epoch training increases, the classifier's output accuracy (%) gradually increases. The four classifier models require an average of lower than 240 s (lower than 4 min) CPU time to complete the training and testing tasks, as seen in Table 6. Then, the trained and untrained feature patterns are randomly selected, and the accuracy (%) of the four classifier models is tested by performing 10-fold cross-validation (K_f = 10). Table 7 shows the overall cross-validation results. Figure 7a indicates that the accuracy (%) of Models #2 and #3 can be improved over 600 epochs of training. By comparison, the accuracy of Models #1 and #4 can be improved over 200–400 epochs, after which it converges, and the accuracy (%) of classification approaches the maximum. The training convergence curve of the classifier is shown in Figure 7b. The accuracy (%) of the four models may reach larger than 95%. To shorten the classifier's design cycle and reduce the memory requirements for storing classifier parameters, we recommend using the architectures of Models #1 and #4 to establish and implement the multilayer CNN-based classifiers.

Table 6. Comparisons of the training CPU time for different models of multilayer CNN-based classifier.

Model	1st Convolutional Layer	2nd Convolutional Layer	2nd Pooling Layer	3rd Convolutional Layer	3rd Pooling Layer	Classification Layer (Fully Connecting Network)	Average Training Time (s)
1	3 × 3, 2	3 × 3, 16	2 × 2, 16	3 × 3, 16	2 × 2, 16	625 input nodes, 168 1st hidden nodes, 64 2nd hidden nodes, 2 output nodes	<280
2		3 × 3, 16		5 × 5, 16			<220
3		5 × 5, 16		3 × 3, 16			<240
4		5 × 5, 16		5 × 5, 16			<330

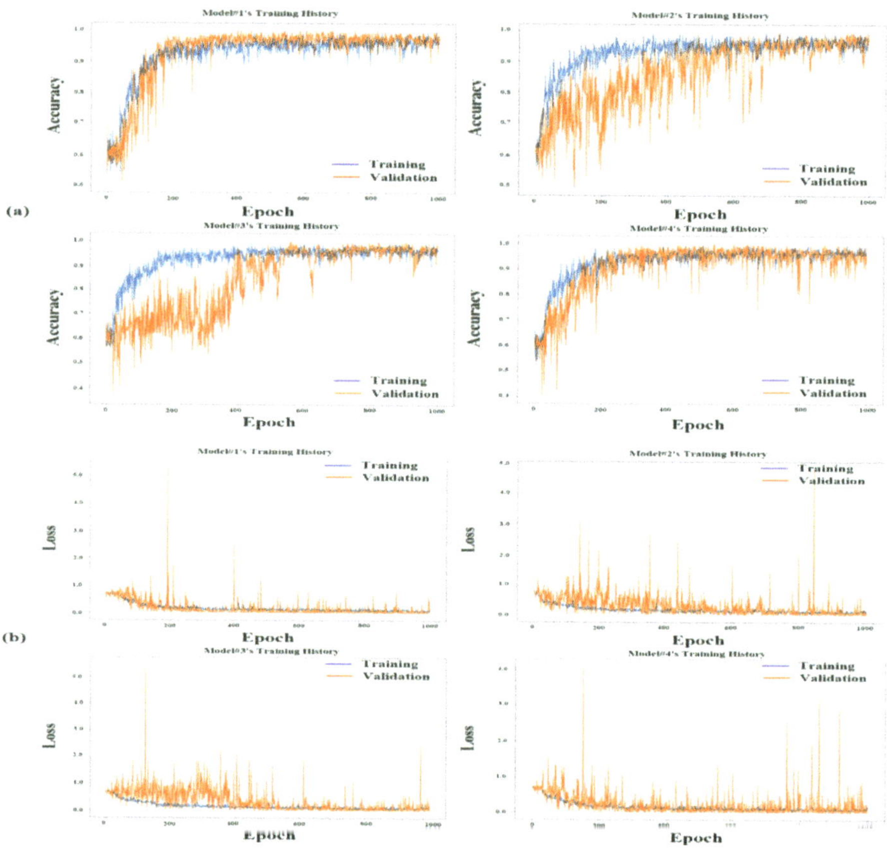

Figure 7. Training history curves of the multilayer CNN-based classifier. (**a**) Training performance test and classification performance validation as seen classification accuracy versus the training epoch and (**b**) classifier training convergence curve as a loss function versus the training epoch.

Table 7. Cross-validation results for different multilayer CNN models (K_f = 10).

Model \ Test Fold	1	2	3	4	5	6	7	8	9	10	Average Accuracy
1	96.14	97.43	98.07	97.96	98.93	98.07	96.35	95.60	96.89	98.28	97.37
2	97.42	98.93	98.28	97.64	99.14	97.21	97.85	98.28	99.14	97.42	98.13
3	96.14	97.64	98.50	98.71	99.14	97.85	97.64	95.06	97.21	99.57	97.75
4	98.93	98.71	96.14	97.32	99.36	98.18	91.74	90.34	97.64	90.02	95.84

Considering the experimental results listed in Table 6, the architecture of Model #1 is selected to establish the screening classifier. After training is completed, 466 untrained feature patterns, including 184 abnormal and 282 normal patterns, are randomly selected from the dataset to validate the performance of the classifier. The experimental results of the classifier produce a visual confusion matrix. The testing result of the abnormal pattern yields TP = 178 and FP = 6, while that of the normal pattern yields TN = 269 and FN = 13; these values can be used as variables in Table 2 to compute the four evaluation indices of the classifier. In this study, precision (%) = 96.74%, recall (%) = 93.19%, F1 score = 0.9493, and accuracy (%) = 95.92%. Precision (%) is the standard for predicting TP, and recall (%) is the true accuracy of TP. Both indicators may be greater than 80%. Recall (%) is also called the positive predictive value (PPV), which is the so-called TP in the detection case. The general PPV index is larger than 80%, which means the proposed classifier has promising predictive performance. The F1 score fuses the indicators of precision (%) and recall (%), and F1 score larger than 0.9000 generally indicates a good classification model. Youden's index (YI) is a fusion evaluation index of sensitivity (Sens) and specificity (Spec) [54], which reflects the performance of the classifier for detecting abnormalities. The larger the YI, the better the performance of the classifier for detection and validation and the greater its authenticity. The testing results show YI = 91.01% (Sens = 93.19%, Spec = 97.82%). Given that all evaluation indicators considered in this work exceed 90%, Model #1 indeed has an architecture that supports good classification accuracy and performance, as seen in the tenfold cross-validation (K_f = 10) for averages of precision (%), recall (%), accuracy (%), and F1 score in Table 8. Hence, we suggest Model #1 to carry out a multilayer CNN-based classifier for automatic breast tumor screening. In addition, as seen in Table 9, we also set 4, 8, 16, and 32 Kernel convolutional windows and 4, 8, 16, and 32 maximum pooling windows in second and third convolutional-pooling layers, respectively, for establishing four models (Models #1–1 to #1–4). With the tenfold cross-validation, trained feature patterns are randomly selected, the average training CPU time of Models #1–1 and #1–2 is less than Model #1–3 with 16 Kernel convolutional windows and 16 maximum pooling windows. It can be seen that Model #1–4 comprises 32 Kernel convolutional windows and 32 maximum pooling windows will increase the average training CPU time and complex computational processes at each cross-validation. With the tenfold cross-validation, untrained feature patterns are also randomly selected, as seen in Tables 10–13, the proposed architecture of multilayer classifier (Model #1–3) has promising classification accuracy and performance in terms of average precision (%), average recall (%), average accuracy (%), and average F1 score. Additionally, the proposed CNN architecture with different convolutional windows in the first convolutional layer, including fractional-order, Sobel (first-order), and Histeq convolutional windows, is used to test the performance of breast tumor screening model. Through the tenfold cross-validation, the CNN classifier with a fractional-order convolutional window in the first convolutional layer, as Model #1 in Table 14, has better classification accuracy (larger than 95%) than Model #2 (larger than 85%) and Model #3 (larger than 90%).

Table 8. Experimental results of K-fold cross-validation (K_f = 10) for the proposed deep-learning-based CNN.

Test Fold	Precision (%)	Recall (%)	Accuracy (%)	F1 Score
1	95.00	96.48	95.60	0.9574
2	94.38	95.92	95.20	0.9514
3	94.82	91.80	93.80	0.9389
4	95.02	96.50	95.60	0.9575
5	96.51	91.68	95.40	0.9577
6	94.09	94.84	94.40	0.9447
7	92.80	97.61	95.00	0.9515
8	92.77	95.85	94.40	0.9429
9	96.46	95.70	96.00	0.9608
10	95.19	95.54	95.00	0.9536
Average	95.19	95.19	95.04	0.9516

Table 9. Comparisons of the training CPU time for multilayer CNN-based classifiers with different numbers of Kernel convolutional windows and maximum pooling windows in second and third convolutional-pooling layers.

Model	1st Convolutional Layer	2nd Convolutional Layer	2nd Pooling Layer	3rd Convolutional Layer	3rd Pooling Layer	Classification Layer (Fully Connecting Network)	Average Training Time (s)
1–1	3 × 3, 2	3 × 3, 4	2 × 2, 4	3 × 3, 4	2 × 2, 4	625 input nodes, 168 1st hidden nodes, 64 2nd hidden nodes, 2 output nodes	<150
1–2		3 × 3, 8	2 × 2, 8	3 × 3, 8	2 × 2, 8		<240
1–3		3 × 3, 16	2 × 2, 16	3 × 3, 16	2 × 2, 16		<280
1–4		3 × 3, 32	2 × 2, 32	3 × 3, 32	2 × 2, 32		<330

Table 10. Experimental results of K-fold cross-validation (K_f = 10) for Model #1–1 with 4 Kernel convolutional windows and 4 maximum pooling windows in each convolutional-pooling layer.

Test Fold	Precision (%)	Recall (%)	Accuracy (%)	F1 Score
1	85.30	84.70	87.80	0.8640
2	87.10	93.00	91.20	0.9000
3	82.80	93.40	89.00	0.8760
4	86.10	91.50	93.20	0.9200
5	84.00	92.00	89.20	0.8780
6	93.20	84.80	91.08	0.8880
7	91.20	95.70	95.40	0.9480
8	90.10	94.80	93.60	0.9260
9	92.40	86.70	91.40	0.8950
10	97.20	99.10	98.40	0.9810
Average	88.94	91.57	92.30	0.9076

Table 11. Experimental results of K-fold cross-validation (K_f = 10) for Model #1–2 with 8 Kernel convolutional windows and 8 maximum pooling windows in each convolutional-pooling layer.

Test Fold	Precision (%)	Recall (%)	Accuracy (%)	F1 Score
1	96.60	95.70	96.80	0.9620
2	94.80	95.60	96.00	0.9530
3	96.10	93.00	95.40	0.9450
4	88.30	96.20	93.00	0.9210
5	90.10	95.30	93.60	0.9260
6	92.10	93.40	93.80	0.9270

Table 11. *Cont.*

Test Fold	Precision (%)	Recall (%)	Accuracy (%)	F1 Score
7	93.40	93.80	94.70	0.9360
8	92.30	93.00	93.80	0.9270
9	94.10	97.60	96.40	0.9580
10	90.30	96.70	94.20	0.9340
Average	92.81	95.03	94.77	0.9389

Table 12. Experimental results of K-fold cross-validation (K_f = 10) for Model #1–3 with 16 Kernel convolutional windows and 16 maximum pooling windows in each convolutional-pooling layer.

Test Fold	Precision (%)	Recall (%)	Accuracy (%)	F1 Score
1	96.70	96.20	97.00	0.9640
2	97.60	95.30	96.60	0.9570
3	95.10	93.40	95.40	0.9420
4	97.10	94.00	96.20	0.9540
5	97.20	97.10	97.60	0.9680
6	93.00	94.00	94.40	0.9340
7	95.60	92.40	95.00	0.9400
8	98.00	97.20	98.10	0.9760
9	96.50	95.70	96.00	0.9610
10	95.20	95.50	95.00	0.9540
Average	96.30	95.04	95.93	0.9553

Table 13. Experimental results of K-fold cross-validation (K_f = 10) for Model #1–4 with 32 Kernel convolutional windows and 32 maximum pooling windows in each convolutional-pooling layer.

Test Fold	Precision (%)	Recall (%)	Accuracy (%)	F1 Score
1	96.70	96.20	97.00	0.9640
2	99.00	96.20	99.00	0.9760
3	90.80	93.40	93.20	0.9210
4	95.60	93.40	95.40	0.9230
5	99.50	97.60	98.80	0.9860
6	95.30	95.70	96.20	0.9550
7	66.00	95.60	77.20	0.7800
8	96.70	98.60	98.00	0.9770
9	97.40	91.50	95.40	0.9440
10	99.00	94.80	97.40	0.9690
Average	93.60	95.30	94.60	0.9395

Table 14. Comparisons of average accuracy (%) for the CNN-based classifier with different convolutional windows in first to third convolutional layers.

Model	First Convolutional Window	Second and Third Convolutional Window	Second and Third Pooling Window	Classification Layer	Average Accuracy
1	Fractional-Order Convolutional Window (2)	Kernel convolution Window (16)	Maximum Pooling Window (16)	One Input Layer, Two Hidden Layers, One Output Layer	>95%
2	Sobel (First-Order) Convolutional Window (2)	Kernel convolution Window (16)	Maximum Pooling Window (16)		>85%
3	Histeq Convolutional Window (2)	Kernel convolution Window (16)	Maximum Pooling Window (16)		>90%

Figure 8 shows the left breast mammogram image of a patient (File Name #mdb315ll [33,34]). In this case study, the right breast mammogram image (File Name #mdb316rl) is normal (background tissue: Dense-glandular (D)), the left breast has a benign tumor (B), the center coordinates of the tumor are (1900, 317), the background tissue is D, and the circumscribed masses are labeled (CIRC) [33,34]. In this study, using the automatic screening system developed [55], as per the pre-selected priority order of screening (①→⑥), the sequence of ROI blocks is shown in Figure 8, and the automatic screening results show four TP (①,②,③, and⑥) and two TNs (④and⑤). In this case, the large tumor spans four ROI blocks①–③and⑥. Therefore, the screening results show TP for identifying a possible breast tumor, the reliability of the classifier output judged to be abnormal is larger than 0.50, and the abnormality is flagged by a red message. The screening system can be switched to manual mode. Similar to the automatic screening results, the four ROI blocks①–③and⑥can be manually circled, screenshotted, and stored in the queue in the order of manual screenshots. The classifier automatically performs image recognition in sequence, and the corresponding recognition results and messages are returned so that the clinician can confirm the possible tumor conditions.

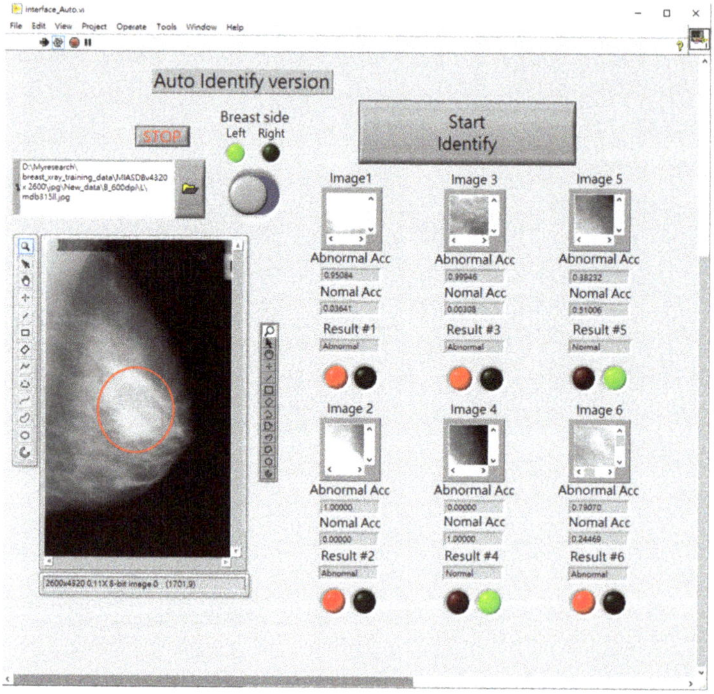

Figure 8. Human–machine interface of the computer assistive system and its automatic screening results.

3.4. Discussion

This study designs a mammography classification method incorporating a multilayer CNN-based classifier for automatic breast tumor screening in clinical applications. The proposed classifier algorithm is implemented in the LabVIEW 2019 (NI^{TM}) software, MATLAB Script tools, and open-source Tensorflow platform (Version 1.9.0) [28] and integrated into a computer assistive system with the automatic and manual feature extraction and breast tumor screening modes. The fractional-order convolutional layer and two convolutional-pooling layers allow the image enhancement and sharpening of the possible tumor edges, contours, and shapes via one fractional-order and two kernel convolutional processes in the feature patterns. Through a series of convolution and pooling processes at different scales and different dimensions, the classifier can obtain nonlinearity feature representation from low-level features to high-level information [29]. Then, with the specific bounding boxes (automatic or manual mode) for ROI extraction, enhanced feature patterns can then be distinguished for further breast tumor screening by the multilayer classifier in the classification layer. A gradient-descent optimization method, namely, the ADAM algorithm, is used in the back-propagation process to adjust the network weighted parameters in the classification layer. With K-fold (K_f = 10) cross-validation, the 466 randomly selected untrained feature patterns for each test fold, the proposed multilayer CNN-based classifier, has high recall (%), precision (%), accuracy (%), and F1 scores for screening abnormalities in both right and left breasts. Experimental results show that the proposed multilayer CNN model offers image enhancement, feature extraction, automatic screening capability, and higher average accuracy (larger than 95%) for separating the normal condition from the possible tumor classes. It has been observed from previous literature [3–7,10,56] that multilayer CNNs comprised several convolutional-pooling layers and a fully connecting network to establish a classifier for automatic breast tumor screening, and could also be applied for CT, MRI, chest X-ray, and ultrasound image processes, such as image classification and segmentation in clinical applications [19,23,28,35,36,51,55]. The combination of a cascade of deep learning and a fully connecting networks is also carried out by a multilayer CNN-based classifier, and a decision scheme [56]. For the screened suspicious region on mammograms, the cascade of the deep-learning method had 98% sensitivity and 90% specificity on the SuReMapp (Suspicious Region Detection on Mammogram from PP) dataset [56], and 94% sensitivity and 91% specificity on the mini-MIAS dataset [56]. This CNN-based multilayer classifier could extract multi-scale feature patterns, and increase the depth and width feature patterns by using multi-convolutional-pooling processes, which had an overall increase in accuracy. However, excessive multi-convolutional processes would completely lead to a loss of the internal data about the position and the orientation of the desired object, and an excessive multi-pooling processing would lose valuable information relating to the spatial relationships between features; thus, many processes were required to perform with GPU hardware for complex computational processes. Hence, the proposed optimal multilayer CNN architecture contained 2D spatial information in the fractional-order convolutional layer (with two fractional-order convolutional windows), and continuously enhanced the features with two-round convolutional-pooling processes (with 16 Kernel convolutional windows and 16 maximum pooling windows), which could extract the desired features at different scales and different levels. Thus, in comparison with the other deep-learning methods, the proposed multilayer classifier exhibited promising results for the desired medical diagnostic purpose. Hence, we have some advantages for the proposed CNN-based classifier, as follows:

- The ROI extraction, image enhancement, and feature classification tasks are integrated into one learning model;
- The fractional-order convolutional process with fractional-order parameter, v = 0.30–0.40, is used to extract the tumor edges in the first convolutional layer; subsequently, two kernel convolution processes are used to extract the tumor shapes;
- The ADAM algorithm is easy to implement and operate with large datasets and parameter adjustment;
- The proposed CNN-based classifier has better classification accuracy than the CNN architecture with Sobel and Histeq convolutional windows in the first convolutional layer.

4. Conclusions

The proposed CNN architecture had better learning ability for complex feature patterns in massive-sized training datasets, and also had more promising classifier performance than traditional CNN-based classifiers and a cascade of deep-learning-based classifiers. Through experimental test and validation, we suggested optimal architecture for a simplified and established multilayer CNN-based classifier, which consisted of a fractional-order convolutional layer, two Kernel convolutional-pooling layers, and a classification layer. Hence, this optimal CNN-based classifier could replace manual screening for tasks requiring specific expertise and experience for medical image examination, which could

26. Zeiler, M.D.; Fergus, R. Visualizing and understanding convolutional networks. In *Lecture Notes in Computer Science;* 8689 LNCS; Springer: Cham, Switzerland, 2014; pp. 818–833.
27. Deng, J.; Dong, W.; Socher, R.; Li, L.-J.; Li, K.; Li, F.-F. ImageNet: A large-scale hierarchical image database. In Proceedings of the 2009 IEEE Conference on Computer Vision and Pattern Recognition, Miami, FL, USA, 20–25 June 2009; pp. 248–255.
28. Li, Y.-C.; Shen, T.-Y.; Chen, C.-C.; Chang, W.-T.; Lee, P.-Y.; Huang, C.-C. Automatic detection of atherosclerotic plaque and calcification from intravascular ultrasound Images by using deep convolutional neural networks. *IEEE Trans. Ultrason. Ferroelectr. Freq. Control* **2021**, *68*, 1762–1772. [CrossRef] [PubMed]
29. Gu, J.; Wang, Z.; Kuen, J.; Ma, L.; Shahroudy, A.; Shuai, B.; Liu, T.; Wang, X.; Wang, G.; Cai, J.; et al. Recent advances in convolutional neural network. *Pattern Recognit.* **2018**, *77*, 354–377. [CrossRef]
30. Kiranyaz, S.; Avci, O.; Abdeljaber, O.; Ince, T.; Gabbouj, M.; Inman, D.J. 1D convolutional neural networks and applications: A survey. *Mech. Syst. Signal Process.* **2021**, *151*, 107s398. [CrossRef]
31. Kingma, D.P.; Ba, J. Adam: A method for stochastic optimization. In Proceedings of the 3rd International Conference for Learning Representations, San Diego, CA, USA, 7–9 May 2015.
32. Ma, J.; Yarats, D. Quasi-hyperbolic momentum and Adam for deep learning. In Proceedings of the International Conference on Learning Representations, New Orleans, LA, USA, 6–9 May 2019. [CrossRef]
33. Pilot European Image Processing Archive. The Mini-MIAS Database of Mammograms. 2012. Available online: http://peipa.essex.ac.uk/pix/mias/ (accessed on 1 January 2022).
34. Mammographic Image Analysis Society (MIAS). Database v1.21. 2019. Available online: https://www.repository.cam.ac.uk/handle/1810/250394 (accessed on 1 January 2022).
35. Wu, J.-X.; Chen, P.-Y.; Li, C.-M.; Kuo, Y.-C.; Pai, N.-S.; Lin, C.-H. Multilayer fractional-order machine vision classifier for rapid typical lung diseases screening on digital chest X-ray images. *IEEE Access* **2020**, *8*, 105886–105902. [CrossRef]
36. Lin, C.-H.; Wu, J.-X.; Li, C.-M.; Chen, P.-Y.; Pai, N.-S.; Kuo, Y.-C. Enhancement of chest X-ray images to improve screening accuracy rate using iterated function system and multilayer fractional-order machine learning classifier. *IEEE Photonics J.* **2020**, *12*, 1–19. [CrossRef]
37. Pu, Y.-F.; Zhou, J.-L.; Yuan, X. Fractional differential mask: A fractional differential-based approach for multiscale texture enhancement. *IEEE Trans. Image Process.* **2010**, *19*, 491–511.
38. Zhang, Y.; Pu, Y.-F.; Zhou, J.-L. Construction of fractional differential masks based on Riemann-Liouville definition. *J. Comput. Inf. Syst.* **2010**, *6*, 3191–3199.
39. Thanh, D.N.H.; Kalavathi, P.; Thanh, L.T.; Prasath, V.B.S. Chest X-ray image denoising using Nesterov optimization method with total variation regularization. *Procedia Comput. Sci.* **2020**, *171*, 1961–1969. [CrossRef]
40. Ba, L.J.; Frey, B. Adaptive dropout for training deep neural networks. *Adv. Neural Inf. Process. Syst.* **2013**, *26*, 1–9.
41. Clevert, D.; Unterthiner, T.; Hochreite, S. Fast and accurate deep network learning by exponential linear units (ELUs). In Proceedings of the 4th International Conference on Learning Representations, San Juan, Puerto Rico, 2–4 May 2016.
42. Hendrycks, D.; Gimpel, K. Gaussian Error Linear Units (GELUs). *arXiv* **2016**, arXiv:1606.08415.
43. Boer, P.; Kroese, D.P.; Mannor, S.; Rubinstein, R.Y. A tutorial on the cross-entropy method. *Ann. Oper. Res.* **2005**, *134*, 19–67. [CrossRef]
44. Rubinstein, R.Y.; Kroese, D.P. *The Cross-Entropy Method: A Unified Approach to Combinatorial Optimization, Monte-Carlo Simulation, and Machine Learning*; Information Science and Statistics; Springer: New York, NY, USA, 2004; pp. 1–47.
45. Ho, Y.; Wookey, S. The real-world-weight cross- entropy loss function: Modeling the costs of mislabeling. *IEEE Access* **2019**, *8*, 4806–4813. [CrossRef]
46. Chen, X.; Liu, S.; Sun, R.; Hong, M. On the convergence of a class of ADAM-type algorithms for non-convex optimization. *arXiv* **2018**, arXiv:1808.02941.
47. Syntax: Edge, 1994–2021 Years. Available online: https://www.mathworks.com/help/images/ref/edge.html (accessed on 1 January 2022).
48. Chicco, D.; Jurman, G. The advantages of the Matthews correlation coefficient (MCC) over F1 score and accuracy in binary classification evaluation. *BMC Genom.* **2020**, *21*, 6. [CrossRef]
49. Syntax: Histeq, 1994–2021 Years. Available online: https://www.mathworks.com/help/images/ref/histeq.html (accessed on 1 January 2022).
50. Syntax: Imadjust, 1994–2021 Years. Available online: https://www.mathworks.com/help/images/ref/imadjust.html (accessed on 1 January 2022).
51. Wu, J.-X.; Liu, H.-C.; Chen, P.-Y.; Lin, C.-H.; Chou, Y.-H.; Shung, K.K. Enhancement of ARFI-VTI elastography images in order to preliminary rapid screening of benign and malignant breast tumors using multilayer fractional-order machine vision classifier. *IEEE Access* **2020**, *8*, 164222–164237. [CrossRef]
52. Valenzuela, G.; Laimes, R.; Chavez, I.; Salazar, C.; Bellido, E.G.; Tirado, I.; Pinto, J.; Guerrero, J.; Lavarello, R.J. In vivo diagnosis of metastasis in cervical lymph nodes using backscatter coefficient. In Proceedings of the 2018 IEEE International Ultrasonics Symposium (IUS), Kobe, Japan, 22–25 October 2018.
53. Chansong, D.; Supratid, S. Impacts of Kernel size on different resized images in object recognition based on convolutional neural network. In Proceedings of the 2021 9th International Electrical Engineering Congress (iEECON), Pattaya, Thailand, 10–12 March 2021.

54. Sidek, K.A.; Khalil, I.; Jelinek, H.F. ECG biometric with abnormal cardiac conditions in remote monitoring system. *IEEE Trans. Syst. Man Cybern. Syst.* **2014**, *44*, 1498–1509. [CrossRef]
55. Zhang, X.-H. A Convolutional Neural Network Assisted Fast Tumor Screening System Based on Fractional-Order Image Enhancement: The Case of Breast X-ray Medical Imaging. Master's Thesis, Department of Electrical Engineering, National Chin-Yi University of Technology, Taichung City, Taiwan, July 2021.
56. Bruno, A.; Ardizzone, E.; Vitabile, S.; Midiri, M. A novel solution based on scale invariant feature transform descriptors and deep learning for the detection of suspicious regions in mammogram images. *J. Med. Signals Sens.* **2020**, *10*, 158–173.

applied sciences

MDPI

Article

The Fusion of MRI and CT Medical Images Using Variational Mode Decomposition

Srinivasu Polinati [1,2], Durga Prasad Bavirisetti [3], Kandala N V P S Rajesh [4], Ganesh R Naik [5,*] and Ravindra Dhuli [6,*]

1 School of Electronics Engineering, VIT University, Vellore 632014, India; srinivasu.polinati@gmail.com
2 Department of ECE, Vignan's Institute of Engineering for Women, Visakhapatnam 530046, India
3 School of Computing Science and Engineering, VIT Bhopal, Bhopal 466114, India; durga.prasad@vitbhopal.ac.in
4 Department of ECE, Gayatri Vidya Parishad College of Engineering, Visakhapatnam 530048, India; kandala.rajesh2014@gmail.com
5 Adelaide Institute for Sleep Health, Flinders University, Bedford Park, SA 5042, Australia
6 School of Electronics Engineering, VIT-AP University, Vijayawada 522237, India
* Correspondence: ganesh.naik@flinders.edu.au (G.R.N.); ravindra.d@vitap.ac.in (R.D.)

Citation: Polinati, S.; Bavirisetti, D.P.; Rajesh, K.N.V.P.S.; Naik, G.R.; Dhuli, R. The Fusion of MRI and CT Medical Images Using Variational Mode Decomposition. *Appl. Sci.* **2021**, *11*, 10975. https://doi.org/10.3390/app112210975

Academic Editors: Michalis Vrigkas and Antonio Fernández

Received: 21 September 2021
Accepted: 16 November 2021
Published: 19 November 2021

Abstract: In medical image processing, magnetic resonance imaging (MRI) and computed tomography (CT) modalities are widely used to extract soft and hard tissue information, respectively. However, with the help of a single modality, it is very challenging to extract the required pathological features to identify suspicious tissue details. Several medical image fusion methods have attempted to combine complementary information from MRI and CT to address the issue mentioned earlier over the past few decades. However, existing methods have their advantages and drawbacks. In this work, we propose a new multimodal medical image fusion approach based on variational mode decomposition (VMD) and local energy maxima (LEM). With the help of VMD, we decompose source images into several intrinsic mode functions (IMFs) to effectively extract edge details by avoiding boundary distortions. LEM is employed to carefully combine the IMFs based on the local information, which plays a crucial role in the fused image quality by preserving the appropriate spatial information. The proposed method's performance is evaluated using various subjective and objective measures. The experimental analysis shows that the proposed method gives promising results compared to other existing and well-received fusion methods.

Keywords: MRI; CT; Image fusion; intrinsic mode functions (IMFs); LEM; VMD

1. Introduction

Medical image analysis plays a crucial role in clinical assessment. However, the success rate of the diagnosis depends upon the visual quality and the information present in medical images [1]. In real-world medical imaging, denoising [2,3] or texture information processing [4,5] is a necessary preprocessing step to improve the fused image's visual quality further.

Nowadays, several imaging modalities are available to capture specific medical information of a given organ [6–8]. X-ray, MRI, CT, positron emission tomography (PET), and single-photon emission computed tomography (SPECT) of a human brain displayed in Figure 1 are crucial medical imaging modalities among them. For example, the magnetic resonance imaging (MRI) modality captures the anatomical information of the soft tissue. In contrast, computed tomography (CT) significantly provides hard tissue information such as bones structures and tumors [8]. Moreover, for clinical needs, the information provided by a single modality may not be sufficient, especially during the diagnosis of diseases [9]. The image fusion mechanism can effectively address this problem, enhancing the information by combining the complementary details provided by two or more modalities into a single image.

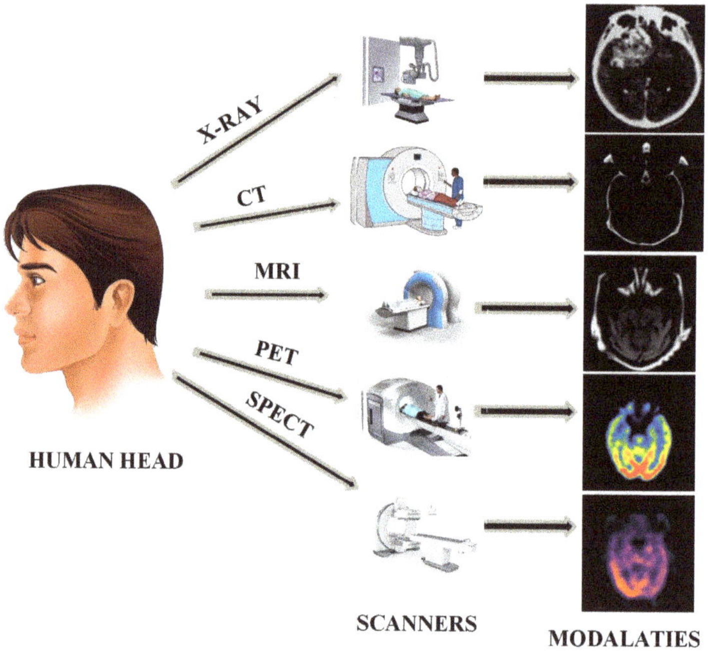

Figure 1. Illustration of the classification of different medical brain imaging modalities.

Image fusion can be categorized into spatial and transform domain techniques [10]. In spatial domain methods, the fusion takes place between the pixels of the source images directly. The maximum, minimum, average, weighted average and PCA are examples of the spatial domain fusion methods, which are easy to implement and computationally efficient. Direct pixel-based fusion methods use a weighted pixel of input images to form a fused image [11]. The activity level of the pixels determines these weights. In the literature, various machine learning methods such as neural networks and support vector machines (SVM) are also used to select the pixels with the highest activity [12,13]. In [14], an iterative block-level fusion method is proposed. First, the source images are decomposed into small square blocks, and PCA is computed on those blocks. Next, weights are found using the average of the PCA components. Finally, a maximum average mutual information fusion rule is employed for the final blending of input images. In [15], a pixel-level image fusion method is proposed using PCA. Here, the first PCA components from both the input images are multiplied individually, and those weighted images are added for fusion. However, these methods might exhibit spatial color, information loss, and brightness distortions [16,17].

Image fusion methods based on the transform domain techniques are receiving much consideration [18]. Pyramid [19], wavelet [20], and multi-resolution singular value decomposition (MSVD) are examples of traditional methods [21] in this category. However, transform domain fusion methods have a few drawbacks [18]. For example, most pyramid methods suffer from blocking artifacts and a loss of source information, even producing artifacts around edges [22]. Wavelets suffer shift sensitivity, poor directionality, an absence of phase information, poor performance at edges and texture regions, and produce artifacts around edges because of the shift-variant nature [22]. Despite the reliable quantification results, MSVD fusion methods might result in poor visual quality [23].

To address the issues mentioned above, other transform domain fusion techniques such as À Trous wavelet transform (ATWT), curvelet transform (CVT), and ridgelet transform are suggested in [24]. These methods provide better results concerning the visual aspect, preserving spatial and spectral information. Nevertheless, these techniques suffer from artifacts around the edges in the fused image [25].

In [26], a new pixel-level image fusion approach using convolutional sparsity-based morphological component analysis (CS-MCA) is introduced. This method achieves sparse representation by combining MCA and sparse convolutional representation into the unified optimization method. This approach might suffer from a spatial consistency problem, resulting in the degradation of spatial details [27]. An NSST-based fusion scheme is proposed in [28].This approach used a blend of NSST with weighted local energy (WLE) and a weighted sum of eight- neighborhood-based modified Laplacian (WSEML) to integrate MRI and CT images. However, this method is a non-adaptive approach. A summary of different types of image fusion methods, their advantages and drawbacks are tabulated in Table 1.

Table 1. Brief summary of the image fusion methods.

Image Fusion Types		Fusion Methods	Advantages	Drawbacks
Spatial domain		Average, minimum, maximum, morphological operators [11], Principal Component Analysis (PCA) [14], Independent Component Analysis (ICA) [29]	Easy to implement. Computationally efficient	Reduces the contrast, produces brightness or color distortions. May give desirable results for a few fusion datasets.
Transform domain	Pyramidal methods	Contrast Pyramid [30], Ratio of the low-pass pyramid [31], Laplacian [19]	Provides spectral information	May produce artifacts around edges. Suffer from blocking artifacts
	Wavelet transform	Discrete wavelet transform (DWT) [15], Shift invariant discrete wavelet transform (SIDWT) [32], Dual-tree complex wavelet transform (DcxDWT) [20]	Provides directional information	May produce artifacts around edges because of shift variant nature. Computationally expensive and demands large memory.
	Multiscale geometric analysis (MGA)	Curvelet [24], Contourlet [33], Shearlet [34], Nonsubsampled Shearlet transform (NSST) [28]	Provides the edges and texture region	Loss in texture parts, high memory requirement, demands high run time.

An adaptive transform-domain fusion technique might provide a better solution to the challenges mentioned above. In these fusion approaches, the basis function of the transform technique depends on the source image's characteristics. With the help of adaptive wavelets, the image's crucial features can be highlighted, which helps in the fusion process. Hence, adaptive wavelets turned out to be a preferable representation compared to standard wavelets. Similar works based on VMD decomposition-based techniques can be found in [35,36]. However, this paper proposes a new adaptive multimodal image fusion strategy based on the combination of variational mode decomposition (VMD) and local energy maxima (LEM) to address the challenges mentioned above. The highlights of the proposed method are as follows:

1. VMD is an adaptive decomposition scheme that decomposes the images as band-limited sub-bands called intrinsic mode functions (IMFs) without introducing boundary distortions and mode-mixing problems. Indeed, the band-limited sub-bands characterize the edge and line features of source images. This decomposition technique can effectively extract the image features from the other transform methods such as wavelet transform

(WT), bi-dimensional empirical mode decomposition (BEMD), and empirical wavelet transform (EWT);

2. The LEM fusion rule extracts the local information from decomposed modes corresponding to two source images pixel by pixel using a windowing operation (3 × 3) and then measures the maximum information value. Hence, using the LEM fusion rule, we can preserve the required complementary visual, edge, and texture information in the IMFs;

3. The proposed approach aims to preserve the information and details of both MRI and CT images into the fused image using VMD and LEM. From visual perception and objective assessment of the fusion results, it is evident that our new image fusion method accomplishes good performance over other existing fusion methods.

The remainder of the paper is arranged as follows: The proposed framework and its mathematical representation are presented in Section 2. The detailed analysis of the simulation results and necessary discussion is presented in Section 3. A final note on the proposed method and future directions is given in Section 4.

2. Proposed Methodology

Our proposed work aims to integrate the details of the soft tissue and dense bone structure provided by MRI and CT medical imaging technologies into a unique image. For this, we have proposed a multimodal medical image fusion based on a blend of VMD and LEM, as shown in Figure 2.

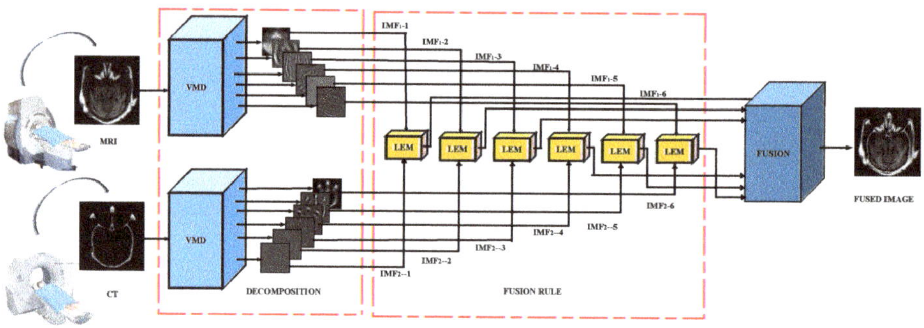

Figure 2. Proposed MRI-CT medical image fusion scheme.

The main steps involved in our fusion methodology are:

A. VMD-based image decomposition;
B. A fusion strategy depending on the LEM;
C. Synthesizing the fused image.

A. VMD-Based Image Decomposition

The traditional decomposition approaches, such as wavelets [37,38], BEMD [39], and EWT [40], suffer from various problems such as boundary distortions and mode-mixing. With these issues, we may fail to achieve an appropriate fusion result. To address these problems, we employed VMD [41], a robust adaptive decomposition approach, highlighting meaningful details in the form of sub-images.

The VMD finds applications in image denoising [42] and texture decomposition [43]. VMD is a non-stationary and adaptive signal processing technique. Unlike EMD and its variants, VMD is not a recursive analysis approach, and it decomposes the signal/image into bandlimited sub-bands based on its frequency content. This work uses VMD to obtain distinct and significant IMFs from the source images (MRI and CT). The derived IMFs reduce mode-mixing and boundary distortions, which are the major concerns in the above mentioned transform domain methods. With this VMD decomposition, we can extract prominent edge information. Initially, we decomposed the input images into six IMFs, which are illustrated in Figure 3.

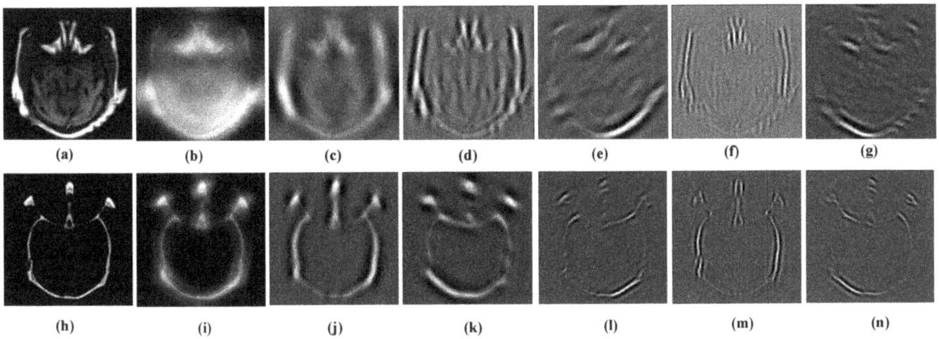

Figure 3. IMFs obtained after VMD decomposition: (**a**) MRI image, (**b**) and (**c–g**) are approximation and detail images of (**a**), respectively. (**h**) CT image, (**i**) and (**j–n**) are approximation and detail images of (**h**), respectively.

From Figure 3, it can be observed that the first IMF ((b) and (i)) captures prominent information from the source images, whereas the remaining IMFs encompass the line and edge information. We can note from Figure 3 that as the mode number increases, the visual details are not significant.

Mathematical Details of VMD:

The main goal of VMD is to subdivide an input signal () into a specific number of sub-bands (IMFs or Modes) (b_l), and each sub-band is bandlimited to specific frequencies in the spectral domain (Fourier domain) by maintaining sparsity. Each of the sub-bands is bandlimited to its center frequencies. VMD involves the following steps to get the bandlimited sub-bands [41]:

1. For each sub-band, its analytical counterpart needs to be computed using Hilbert transform to get the one-sided frequency spectrum;

2. An exponential is used to mix with each mode to shift its frequency spectrum to the baseband;

3. Finally, the bandwidth of the mode estimates using the squared L^2-norm of the gradient. The constrained variational problem can be represented as below.

$$\min_{\{b_l\},\{\omega_l\}} \left\{ \sum_l \|\partial_t \left[\left(\delta(t) + \frac{j}{\pi t} \right) * b_l(t) \right] e^{-j\omega_l t} \|_2^2 \right\} \quad (1)$$
$$\{b_l\} \text{ and } \{w_l\}$$

where l^{th} indicates the lth sub-band and its center frequency, respectively. $\delta(t)$ represents the Dirac distribution, $*$ is the symbol of the convolution.

The constrained problem in Equation (1) is solved using the quadratic penalty term and Lagrangian multipliers λ to make it an unconstrained problem given in Equation (2).

$$L(\{b_l\},\{\omega_l\},\lambda) = \alpha \sum_l \|\partial_t \left[\left(\delta(t) + \frac{j}{\pi t} \right) * b_l(t) \right] e^{-i\omega_l t} \|_2^2 + \|x(t) - \sum_l b_l(t)\|_2^2 + \left\langle \lambda, x(t) - \sum_l b_l(t) \right\rangle \quad (2)$$

where L represents augmented Lagrange matrix function, α is the penalty factor parameter, λ indicates the Lagrange multiplier, and $x(t)$ is the input signal.

Now the solution of Equation (1) can be computed as the saddle point of Equation (2) using the method called an alternating direction method of multipliers (ADMM).

Equation (3) can be further solved using an alternating direction method of multipliers (ADMM) [41]. Finally, the estimate of the lth sub-band is computed as [44]:

$$\hat{b}_l^{n+1}(\omega) = (\hat{x}(\omega) - \sum_{j\neq l} \hat{b}_j(\omega) + \frac{\hat{\lambda}(\omega)}{2}) \frac{1}{1 + 2\alpha(\omega - \omega_l)^2} \tag{3}$$

Similarly, the center frequency is updated as:

$$\omega_l^{n+1} = \frac{\int_0^{\infty} \omega \left| \hat{b}_l(\omega) \right|^2 d\omega}{\int_0^{\infty} \left| \hat{b}_l(\omega) \right|^2 d\omega} \tag{4}$$

In this work, we used the two-dimensional (2D)-VMD [45] method to decompose the MRI and CT images. As stated above, 2D-VMD is a helpful method in extracting useful information such as edges and curves from the source images. Furthermore, VMD is a reliable method to deal with noisy images. Therefore, it can improve the quality of the fusion process even without employing additional preprocessing techniques.

B. Fusion Strategy Depending on LEM

As discussed before, the VMD adaptively decomposes the input images into bandlimited sub-bands called IMFs. Indeed, these IMFs characterize the image features of source images. To highlight and extract relevant features in the fused image, we require appropriate fusion rules. As discussed in Section 1, many fusion rules [46], such as minima, maxima, averaging, and PCA, have been widely explored for this purpose over the past few years. Among them, minima and maxima cause brightness distortions, averaging rule blurs the fused image, and PCA degrades the spectral information [15]. Furthermore, the fusion rules mentioned above may produce low spatial resolution issues [47]. The LEM-based [47] fusion rule is adopted to tackle the issues discussed above in this work.

We have demonstrated the influence of these fusion rules visually in Figure 4 and quantitatively in Table 2. As shown in Figure 4, the VMD with LEM fusion rule achieves visually satisfying results compared to VMD with other fusion rules. Similarly, as shown in Table 2, the fusion metric values calculated over 10 data sets proved the efficacy of the chosen LEM fusion rule.

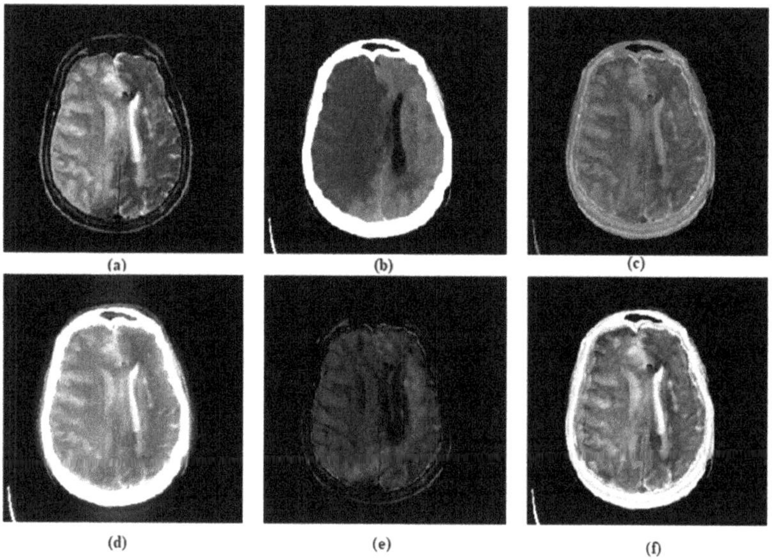

Figure 4. Visual quality analysis of various fusion rules on MRI-CT image pair. (**a**) MRI image, (**b**) CT image, (**c**) VMD-AVG, (**d**) VMD-MAX, (**e**) VMD-MIN (**f**) VMD-LEM.

Table 2. Average quantitative analysis of various fusion rules on 10 pairs of MRI-CT images.

Metrics	Methods			
	VMD-AVG	VMD-MAX	VMD-MIN	VMD-LEM
EI	48.439	58.322	36.487	**71.751**
MI	4.384	4.376	3.486	**4.391**
VIFF	0.335	0.397	0.063	**0.428**
$Q_P^{AB/F}$	0.307	0.356	0.198	**0.443**
SSIM	0.599	0.232	0.563	**0.621**
AG	4.845	5.714	3.735	**6.973**
RMSE	0.0296	**0.005**	0.036	0.020
PSNR	15.926	14.553	15.869	**18.580**

The technical details of the LEM fusion rule are discussed as follows. The principal idea behind using LEM is to extract and preserve vital information with the help of local information constraints from both the images pixel by pixel [47]. The entire process of LEM is described in Algorithm 1.

Algorithm 1

Let us consider the IMFs of the first image as IMFs_A^i, and the sec ond image as IMFs_B^i. The local information $LE_\alpha(x,y)$ of $IMFs_\alpha^i(\alpha = A,B)$ is evaluated using the following steps.
Input : Decomposed modes of images IMFs_A^i, IMFs_B^i.
Output : Enhanced decomposition modes $\mathrm{F}^i{}_{IMFs_{A,B}}(x,y)$.
Step 1 : Calculate the local information $LEM_\alpha(x,y)$ of individual modes $\mathrm{IMFs}_\alpha^i(\alpha = A,B)$

$$\mathrm{LEM}_\alpha(x,y) = \sum_{i=1}^{w}\sum_{j=1}^{w}\left[IMFs_\alpha^i(x+i,y+j)\right]^2 \times W_k(i,j) \quad (5)$$

where, W_k is given by:

$$\mathrm{W}_k = \begin{bmatrix} 1 & 1 & 1 \\ 1 & 1 & 1 \\ 1 & 1 & 1 \end{bmatrix}$$

Step 2 : Choose the maximum value in the local information $LEM_\alpha(x,y)$

$$\mathrm{L}_\alpha(x,y) = \max\{LEM_\alpha(x+i,y+j)|1 \le i,j \le 3\} \quad (6)$$

Step 3: Calculate the binary decision weight maps

$$X_1(x,y) = \begin{cases} 1, \text{ if } L_A(\mathrm{x,y}) > L_B(\mathrm{x,y}) \\ 0, \quad \text{otherwise} \end{cases} \quad (7)$$

$$X_2(x,y) = \begin{cases} 1, \text{ if } L_B(\mathrm{x,y}) > L_A(\mathrm{x,y}) \\ 0, \quad \text{otherwise} \end{cases} \quad (8)$$

Step 4 : Obtain the enhanced decomposition modes $\mathrm{F}^i{}_{IMFs_{A,B}}(x,y)$

$$\mathrm{F}^i{}_{IMFs_{A,B}} = X_1(x,y) \times IMFs_A^i(x,y) + X_2(x,y) \times IMFs_B^i(x,y) \quad (9)$$

C. Synthesizing the Fused Image

We linearly combine all the enhanced IMFs obtained from each LEM fusion rule to construct the fused image. The whole process of the proposed fusion framework is given in Algorithm 2.

Algorithm 2

Input: Image A (MRI), Image B (CT).
Output: The fused image F.
Step 1: Image decomposition using VMD:
Employ VMD on the source images (A and B) to obtain IMF_S which are represented as

$$\begin{aligned} VMD(A) &= \left\{IMFs_A^1, IMFs_A^2 \ldots IMFs_A^i\right\}., i = (1,2,\ldots N); \\ VMD(B) &= \left\{IMFs_B^1, IMFs_B^2 \ldots IMFs_B^i\right\}., i = (1,2,\ldots N) \end{aligned} \quad (10)$$

Step 2: LEM-based image fusion:
(a) Estimate the local information $LEM_\alpha(x,y)$ from each sub − band $IMFs_\alpha^i(\alpha = A, B)$ using Equation (5).
(b) Consider the maximum value $L_\alpha(x,y)$ of $LEM_\alpha(x,y)$ by Equation (6).
(c) Evaluate the binary decision weight maps $X_1(x,y)$, $X_2(x,y)$ with Equations (7) and (8).
(d) Fuse the decomposed modes $F^i{}_{IMFs_{A,B}}(x,y)$ using Equation (9).
Step 3: Reconstruct the fused image by summing all the fused sub-bands obtained from Step 2.

$$F = \sum_{i=1}^{N} F^i{}_\alpha(x,y)\ , i = 1,\ldots N \quad (11)$$

D. Image Fusion Evaluation Metrics

In this paper, we used a few state-of-the-art image fusion metrics to estimate the information contribution of each source image in the fusion process. They are edge intensity (EI) [48], mutual information (MI) [49], visual information fidelity (VIF) [50,51], edge-based similarity measure ($Q_P^{AB/F}$) [52], structural similarity index measure (SSIM) [51,53], average gradient (AG) [54], root mean square error (RMSE) [15], peak signal-to-noise ratio (PSNR) [13,42]. *EI* represents the difference of luminance along the gradient direction in images. *MI* is used to measure the relative information between the source and the fused images. *VIF* estimates the visual information fidelity between the fused and source images depending on the Gaussian mixture model. The edge-based similarity ($Q_P^{AB/F}$) measure will be useful to provide the edge details in the fused image. RMSE computes a difference measure between the reference image and fused image. In this work, the maximum value of RMSE of MRI-fused images and CT-fused images is considered. Similarly, PSNR is also computed. Except for RMSE, the higher values of all these metrics imply better fusion. In the case of the RMSE, the lowest value yields a better result.

3. Results and Discussion

This section presents the experimental setup, results and analysis of the proposed method. First, we explain the experimental setup and methods, followed by data analysis using both qualitative and quantitative methods. Finally, we compare the proposed method with the existing literature for a fair assessment.

The experiments are conducted on a PC with Intel(R) Core (TM) i5-5200U CPU@2.20GHz and RAM 8GB using MATLAB2018b. We have considered a whole-brain atlas website (http://www.med.harvard.edu/AANLIB/home.html, accessed on 1 September 2021) to conduct our experiments. For this purpose, 23 MRI-CT medical image data sets are taken from this database. All these data sets are registered with a resolution of 256 × 256. Image registration [55] is a necessary step prior to image fusion. It is defined as the process of mapping the input images with the help of a reference image. Such mapping aims to match the corresponding images based on specific features to assist in the image fusion process. The database contains various cross-sectional multimodal medical images, such as MRI (T1 and T2 weighted), CT, single-photon emission computed tomography (SPECT), and positron emission tomography (PET).

Furthermore, it has a wide range of brain images ranging from healthy to different brain diseases, including cerebrovascular, neoplastic, degenerative, and infectious diseases.

We have considered 23 pairs of MRI-CT from fatal stroke (cerebrovascular disease) to validate our proposed approach (Supplementary Materials). Interested readers can find more details of this database in [56].

The efficacy of any image fusion algorithm can be verified using subjective (qualitative) and objective (quantitative) analysis. In Section 3.1, we first verified the subjective performance of various fusion algorithms and then performed objective analysis using fusion metrics in Section 3.2.

3.1. Subjective Assessment

Visual results of various MRI and CT fusion methods are shown in Figures 5–7. A good MRI- and CT-fused image should contain both the soft tissue information and dense structure information of the MRI and CT images. We can draw the following observations by examining the visual quality of the four sets of MRI-CT fusion results using various methods.

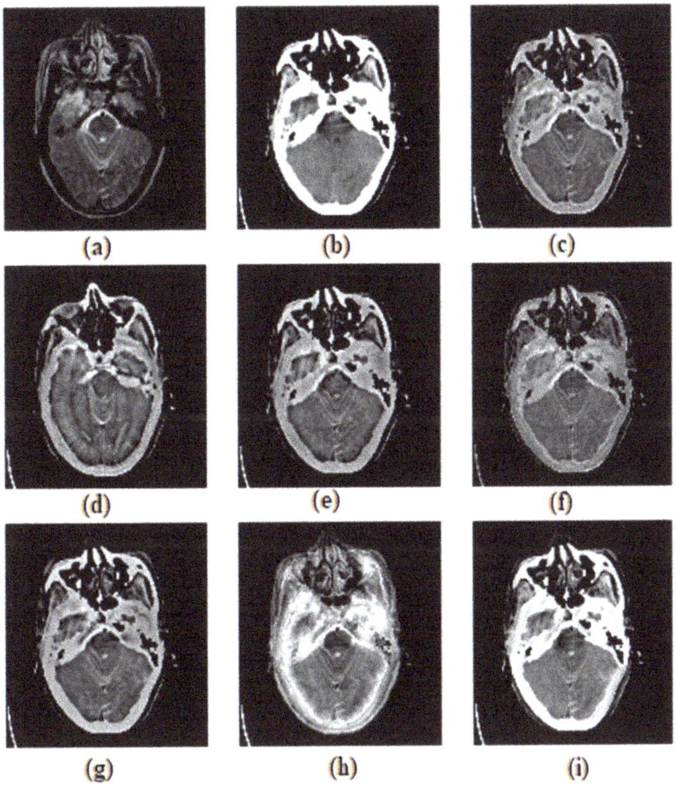

Figure 5. Visual quality analysis of various fusion algorithms for MRI-CT (set-7). (**a**) MRI image, (**b**) CT image, (**c**) ASR, (**d**) CVT, (**e**) DTCWT, (**f**) MSVD, (**g**) CSMCA, (**h**) NSST, (**i**) proposed method.

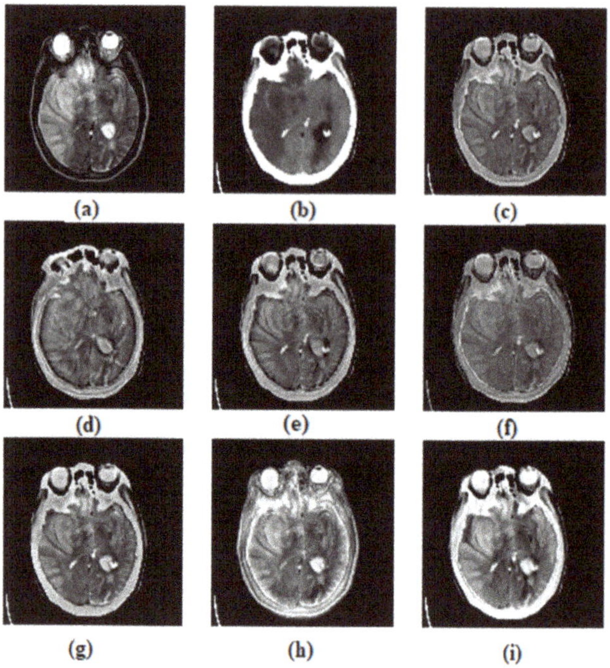

Figure 6. Visual quality analysis of various fusion algorithms for MRI-CT (set-11). (**a**) MRI image, (**b**) CT image, (**c**) ASR, (**d**) CVT, (**e**) DTCWT, (**f**) MSVD, (**g**) CSMCA, (**h**) NSST, (**i**) proposed method.

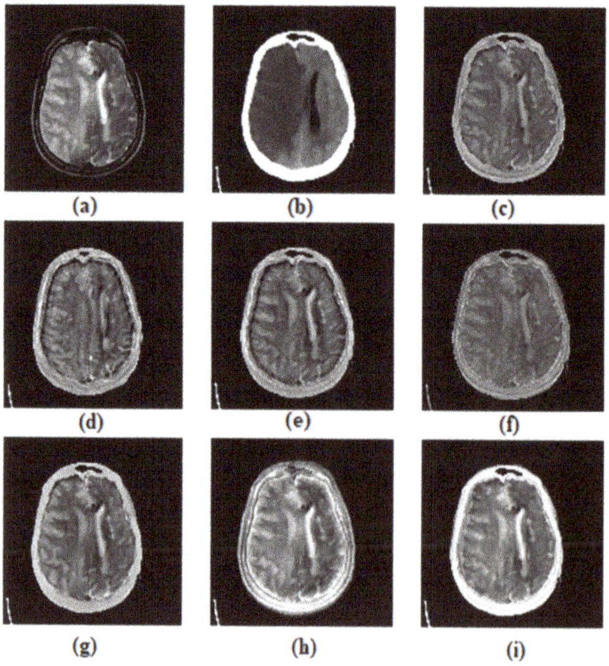

Figure 7. Visual quality analysis of various fusion algorithms for MRI-CT (set-15). (**a**) MRI image, (**b**) CT image, (**c**) ASR, (**d**) CVT, (**e**) DTCWT, (**f**) MSVD, (**g**) CSMCA, (**h**) NSST, (**i**) proposed method.

1. Compared to all the other methods, our proposed algorithm provides a brighter outer region representing the CT image's dense structure;

2. From Figures 5–7, it can be seen that the fused images of methods (c)–(g) are yielding poor contrast;

3. Though the method (h) in all the Figures 5–7 provides better contrast details; still, it is suffering from artifacts, especially in the CT region.

From Figures 5–7, it can be noticed that the ASR method transfers both the CT and MRI information partially with low contrast. Next, coming to the CVT contains more MRI details than the CT. In the DTCWT method, we can find a few fusion artifacts in and around the CT region. Similarly, we can observe information fusion loss in the MSVD method. Compared with the methods mentioned above, CSMCA gives better visual quality, but the overall contrast of the image is reduced. The fused images with the NSST method are visually degraded due to both the fusion loss and artifacts. Overall, our proposed method retains the necessary information from the MRI and CT with minimum fusion losses. The comparison results of the MRI-CT fusion using various methods, including the proposed method on the 23 pairs of fatal stroke images, are shown in Figures 8 and 9.

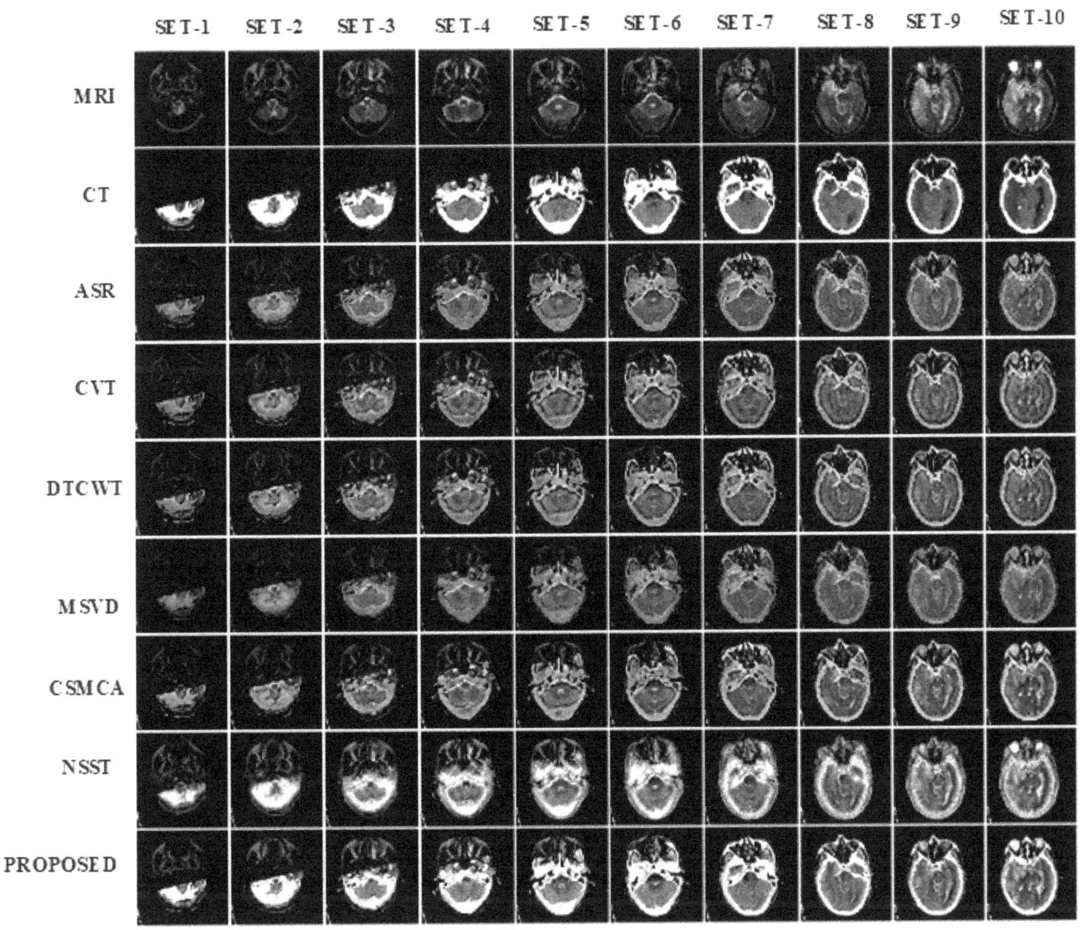

Figure 8. The results of various methods on first 10 pairs of MRI-T images (fatal stroke).

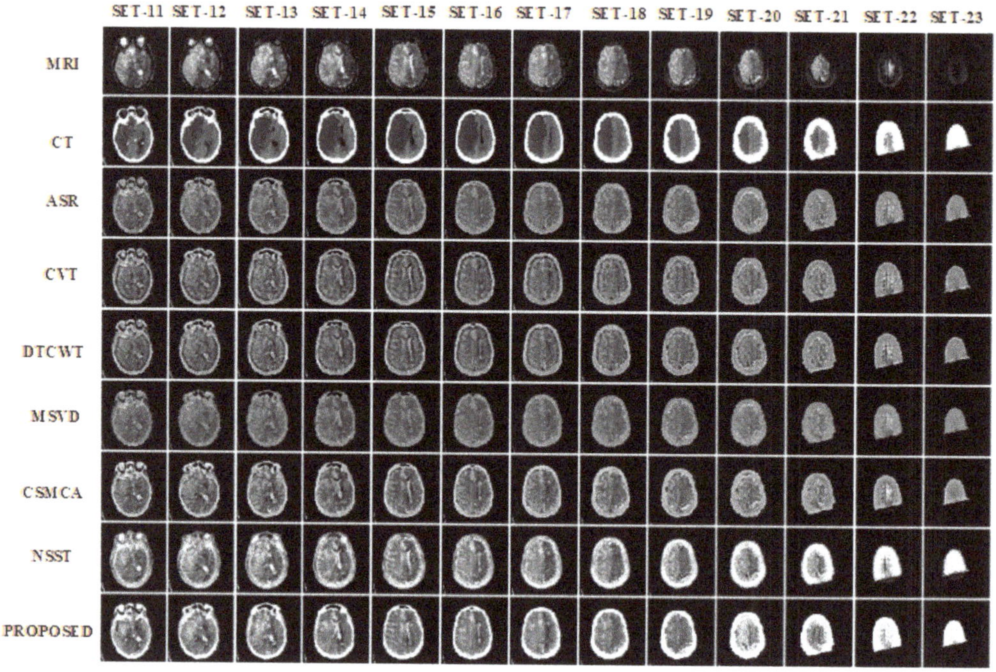

Figure 9. The results of various methods on next 13 pairs of MRI-CT images (fatal stroke).

3.2. Objective Assessment

Here, we assess the fused image quality objectively using fusion metrics. Tables 3–5 demonstrate the objective assessment of the three fatal-stroke images proposed and other existing approaches (sets: 7, 11, and 15) subjectively analyzed earlier. In addition, we have presented the average objective metric scores of all the 23 sets (fatal-stroke)in Table 6. Fusion metrics except for RMSE with the first highest values are highlighted in bold font, and the second-highest values are underlined. The first-lowest value of the RMSE is indicated in bold, and the second-lowest value is underlined. A number within the bracket at the end of the quantitative metric scores represents the rank of the fusion algorithm. In these Tables, the ranking scheme is considered for better quantitative analysis of fusion algorithms.

Table 3. Quantitative analysis of various fusion methods for MRI-CT (set-7).

Metrics	Methods						
	ASR	CVT	DTCWT	MSVD	CSMCA	NSST	Proposed Method
EI	85.184	**91.417 (1)**	88.853	77.183	87.219	81.907	90.390 (2)
MI	**3.948 (2)**	3.548	3.656	3.490	3.811	3.703	**4.079 (1)**
VIFF	0.321	0.290	0.280	0.344 (2)	0.319	0.267	**0.406 (1)**
$Q_P^{AB/F}$	0.535	0.478	0.500	0.427	0.536 (2)	0.373	**0.538 (1)**
SSIM	0.563	0.376	0.499	0.548	0.629 (2)	0.520	**0.697 (1)**
AG	8.561	**9.140 (1)**	8.933	8.332	8.674	8.368	9.008 (2)
RMSE	0.034	0.034	0.034	0.034	0.035	0.027 (2)	**0.020**
PSNR	16.328	16.749	17.166	13.28	17.393 (2)	13.976	**21.342 (1)**

Table 4. Quantitative analysis of the various fusion methods for MRI-CT (set-11).

Metrics	Methods						
	ASR	CVT	DTCWT	MSVD	CSMCA	NSST	Proposed Method
EI	67.026	79.944 (2)	75.086	64.169	70.435	75.318	**80.087 (1)**
MI	4.279	3.904	4.030	4.227	**4.346 (1)**	4.116	4.339 (2)
VIFF	0.272	0.254	0.249	0.286	0.297 (2)	0.241	**0.356 (1)**
$Q_P^{AB/F}$	0.472	0.421	0.435	0.392	**0.481 (1)**	0.421	0.480 (2)
SSIM	0.593	0.276	0.413	0.301	0.537	**0.600 (1)**	0.599 (2)
AG	6.662	7.887 (2)	7.421	6.812	6.877	7.471	**7.980 (1)**
RMSE	0.029	0.029	0.029	0.028	0.029	0.024 (2)	**0.021 (1)**
PSNR	16.857	17.171	17.720	15.804	**17.892 (1)**	13.981	17.794 (2)

Table 5. Quantitative analysis of the state-of-the-art methods for MRI-CT (set-15) dataset.

Metrics	Methods						
	ASR	CVT	DTCWT	MSVD	CSMCA	NSST	Proposed Method
EI	51.347	63.877	58.355	49.732	51.899	65.474 (2)	**65.802 (1)**
MI	4.186	3.878	3.995	4.090	4.284 (2)	4.214	**4.549 (1)**
VIFF	0.356	0.362	0.365	0.348	0.412 (2)	0.340	**0.484 (1)**
$Q_P^{AB/F}$	0.465 (2)	0.418	0.431	0.380	0.461	0.446	**0.478 (1)**
SSIM	0.674 (2)	0.338	0.507	0.417	0.663	0.590	**0.694 (1)**
AG	5.065	6.231	5.719	5.197	5.045	**6.349 (1)**	6.326 (2)
RMSE	0.028	0.029	0.029	0.026	0.028	0.022 (2)	**0.018 (1)**
PSNR	17.396	17.268	17.649	16.392	**18.644 (1)**	14.096	18.024 (2)

Table 6. Average quantitative analysis of the proposed method (23 pairs of MRI-CT) and other state-of-the-art methods.

Metrics	Methods						
	ASR	CVT	DTCWT	MSVD	CSMCA	NSST	Proposed Method
EI	57.800	64.531	61.820	50.850	58.592	62.404	**64.582**
MI	3.666	3.360	3.446	3.694	3.657	3.740	**3.830**
VIFF	0.376	0.362	0.358	0.365	0.401	0.364	**0.498**
$Q_P^{AB/F}$	0.541	0.483	0.500	0.399	0.531	0.439	**0.542**
SSIM	0.651	0.350	0.503	0.614	0.634	0.586	**0.657**
RMSE	0.029	0.029	0.029	0.029	0.029	0.022	**0.020**
AG	5.772	6.390	6.148	5.427	5.771	6.217	**6.412**
PSNR	16.803	16.972	17.242	16.000	17.757	16.021	**20.291**

Comprehensively, the proposed framework is the only approach that occupies the first two ranks for all eight metrics among all the seven methods. It indicates that our method has robust performance (i.e., stable and promising performance) than other existing techniques. Specifically, our approach always remains in the first position on VIFF and RMSE for all four data sets, as shown in Tables 3–5.

Average quantitative analysis of the proposed and other state-of-the-art methods calculated over 23 pairs of MRI-CT (fatal stroke) are presented in Table 6. The proposed method occupied the first position by overperforming other fusion algorithms when average values are considered in fusion metrics.

In general, the consistent performance of any image fusion algorithm in quantitative results is mainly due to the good visual quality of fused images, fusion gain, and less fusion loss and fusion artifacts. We have already seen from the visual result analysis that the proposed method can transfer the source image information into the fused image with less fusion loss and artifacts compared to the other fusion algorithms. It is also evident from the fusion metrics that our method is giving a stable performance.

Hence, we can conclude that the proposed method is promising, stable, and efficient from qualitative and quantitative comparative analysis.

4. Conclusions and Future Scope

We proposed a multi-modal medical image fusion framework with VMD and LEM to fuse MRI and CT medical images in this work. By using an adaptive decomposition technique VMD, significant IMFs are derived from the source images. This decomposition process can preserve some details of source images. However, these details are not sufficient to fulfill the clinical needs of radiologists. Hence, we used a LEM fusion rule to preserve complementary information from IMFs, an essential criterion during medical image diagnosis. All the experiments are evaluated on the Whole Brain Atlas benchmark data sets to analyze the efficacy of the proposed methodology. The experimental results reveal that the proposed framework attained better visual perception. Even objective assessment in terms of average EI (64.582), MI (3.830), VIFF (0.498), $Q_P^{AB/F}$ (0.542), SSIM (0.6574), RMSE (0.020), AG (6.41), and PSNR (20.291) demonstrated quantitative fusion performance better than the existing multi-modal fusion approaches. In the future, we wish to conduct experiments with extensive data that contain images of MRI and CT with different disease information. Additionally, we consider extending this work to both 2D and 3D image clinical applications. Furthermore, we would like to verify the effectiveness of the proposed method for other image fusion applications such as digital photography, remote sensing, battlefield monitoring, and military.

Supplementary Materials: The following are available online at https://www.mdpi.com/article/10.3390/app112210975/s1. The qualitative (Figures S1–S23) and quantitative results (Tables S1–S23) of all the 23 pairs of images used in this work are given in the supplementary material.

Author Contributions: Conceptualization, S.P. and R.D.; methodology, S.P., D.P.B., R.D., K.N.V.P.S.R. and G.R.N.; software, S.P., D.P.B. and K.N.V.P.S.R.; validation, S.P., D.P.B. and K.N.V.P.S.R.; formal analysis, S.P. and G.R.N.; investigation, K.N.V.P.S.R. and G.R.N.; resources, S.P. and R.D.; data curation, K.N.V.P.S.R. and D.P.B.; writing—original draft preparation, S.P., D.P.B. and K.N.V.P.S.R.; writing—review and editing, S.P., D.P.B. and K.N.V.P.S.R.; visualization, S.P., R.D. and G.R.N.; supervision, R.D. and G.R.N.; project administration, K.N.V.P.S.R. All authors have read and agreed to the published version of the manuscript.

Funding: This research received no external funding.

Institutional Review Board Statement: Not applicable.

Informed Consent Statement: Not applicable.

Data Availability Statement: Imaging data can be downloaded from the link: https://www.med.harvard.edu/aanlib/, accessed on 15 November 2021.

Conflicts of Interest: The authors declare no conflict of interest.

References

1. Vishwakarma, A.; Bhuyan, M.K. Image Fusion Using Adjustable Non-subsampled Shearlet Transform. *IEEE Trans. Instrum. Meas.* **2018**, *68*, 3367–3378. [CrossRef]
2. Ouahabi, A. A review of wavelet denoising in medical imaging. In Proceedings of the 2013 8th International Workshop on Systems, Signal Processing and their Applications (WoSSPA), Algiers, Algeria, 12–15 May 2013; IEEE: New York, NY, USA, 2013; pp. 19–26.

3. Ahmed, S.; Messali, Z.; Ouahabi, A.; Trepout, S.; Messaoudi, C.; Marco, S. Nonparametric Denoising Methods Based on Contourlet Transform with Sharp Frequency Localization: Application to Low Exposure Time Electron Microscopy Images. *Entropy* **2015**, *17*, 3461–3478. [CrossRef]
4. Unser, M. Texture classification and segmentation using wavelet frames. *IEEE Trans. Image Process.* **1995**, *4*, 1549–1560. [CrossRef]
5. Meriem, D.; Abdeldjalil, O.; Hadj, B.; Adrian, B.; Denis, K. Discrete wavelet for multifractal texture classification: Application to medical ultrasound imaging. In Proceedings of the 2010 IEEE International Conference on Image Processing, Hong Kong, China, 26–29 September 2010; IEEE: New York, NY, USA, 2010; pp. 637–640.
6. Hatt, C.R.; Jain, A.K.; Parthasarathy, V.; Lang, A.; Raval, A.N. MRI—3D ultrasound—X-ray image fusion with electromagnetic tracking for transendocardial therapeutic injections: In-vitro validation and in-vivo feasibility. *Comput. Med. Imaging Graph.* **2013**, *37*, 162–173. [CrossRef]
7. Labat, V.; Remenieras, J.P.; BouMatar, O.; Ouahabi, A.; Patat, F. Harmonic propagation of finite amplitude sound beams: Experimental determination of the nonlinearity parameter B/A. *Ultrasonics* **2000**, *38*, 292–296. [CrossRef]
8. Dasarathy, B.V. Medical image fusion: A survey of the state of the art. *Inf. Fusion* **2014**, *19*, 4–19. [CrossRef]
9. Zhao, W.; Lu, H. Medical Image Fusion and Denoising with Alternating Sequential Filter and Adaptive Fractional Order Total Variation. *IEEE Trans. Instrum. Meas.* **2017**, *66*, 2283–2294. [CrossRef]
10. El-Gamal, F.E.-Z.A.; Elmogy, M.; Atwan, A. Current trends in medical image registration and fusion. *Egypt. Inf. J.* **2016**, *17*, 99–124. [CrossRef]
11. Li, S.; Kang, X.; Fang, L.; Hu, J.; Yin, H. Pixel-level image fusion: A survey of the state of the art. *Inf. Fusion* **2017**, *33*, 100–112. [CrossRef]
12. Li, S.; Kwok, J.T.; Wang, Y. Multifocus image fusion using artificial neural networks. *Pattern Recognit. Lett.* **2002**, *23*, 985–997. [CrossRef]
13. Li, S.; Kwok, J.-Y.; Tsang, I.-H.; Wang, Y. Fusing Images with Different Focuses Using Support Vector Machines. *IEEE Trans. Neural Netw.* **2004**, *15*, 1555–1561. [CrossRef]
14. Vijayarajan, R.; Muttan, S. Iterative block level principal component averaging medical image fusion. *Optik* **2014**, *125*, 4751–4757. [CrossRef]
15. Naidu, V.; Raol, J. Pixel-level Image Fusion using Wavelets and Principal Component Analysis. *Def. Sci. J.* **2008**, *58*, 338–352. [CrossRef]
16. Singh, S.; Anand, R.S. Multimodal Medical Image Fusion Using Hybrid Layer Decomposition with CNN-Based Feature Mapping and Structural Clustering. *IEEE Trans. Instrum. Meas.* **2020**, *69*, 3855–3865. [CrossRef]
17. Du, J.; Li, W.; Lu, K.; Xiao, B. An overview of multi-modal medical image fusion. *Neurocomputing* **2016**, *215*, 3–20. [CrossRef]
18. Kappala, V.K.; Pradhan, J.; Turuk, A.K.; Silva, V.N.H.; Majhi, S.; Das, S.K. A Point-to-Multi-Point Tracking System for FSO Communication. *IEEE Trans. Instrum. Meas.* **2021**, *70*, 1–10. [CrossRef]
19. Mitianoudis, N.; Stathaki, T. Pixel-based and region-based image fusion schemes using ICA bases. *Inf. Fusion* **2007**, *8*, 131–142. [CrossRef]
20. Toet, A.; van Ruyven, L.J.; Valeton, J.M. Merging Thermal And Visual Images By A Contrast Pyramid. *Opt. Eng.* **1989**, *28*, 287789. [CrossRef]
21. Toet, A. Image fusion by a ratio of low-pass pyramid. *Pattern Recognit. Lett.* **1989**, *9*, 245–253. [CrossRef]
22. Li, X.; Guo, X.; Han, P.; Wang, X.; Li, H.; Luo, T. Laplacian Redecomposition for Multimodal Medical Image Fusion. *IEEE Trans. Instrum. Meas.* **2020**, *69*, 6880–6890. [CrossRef]
23. Li, H.; Manjunath, B.S.; Mitra, S.K. Multisensor Image Fusion Using the Wavelet Transform. *Graph. Model. Image Process.* **1995**, *57*, 235–245. [CrossRef]
24. Lewis, J.J.; O'Callaghan, R.J.; Nikolov, S.G.; Bull, D.R.; Canagarajah, N. Pixel- and region-based image fusion with complex wavelets. *Inf. Fusion* **2007**, *8*, 119–130. [CrossRef]
25. Nencini, F.; Garzelli, A.; Baronti, S.; Alparone, L. Remote sensing image fusion using the curvelet transform. *Inf. Fusion* **2007**, *8*, 143–156. [CrossRef]
26. Yang, L.; Guo, B.L.; Ni, W. Multimodality medical image fusion based on multiscale geometric analysis of contourlet transform. *Neurocomputing* **2008**, *72*, 203–211. [CrossRef]
27. Miao, Q.; Shi, C.; Xu, P.; Yang, M.; Shi, Y. A novel algorithm of image fusion using shearlets. *Opt. Commun.* **2011**, *284*, 1540–1547. [CrossRef]
28. Yin, M.; Liu, X.; Liu, Y.; Chen, X. Medical Image Fusion With Parameter-Adaptive Pulse Coupled-Neural Network in Nonsubsampled Shearlet Transform Domain. *IEEE Trans. Instrum. Meas.* **2018**, *68*, 49–64. [CrossRef]
29. Kirankumar, Y.; Shenbaga Devi, S. Transform-based medical image fusion. *Int. J. Biomed. Eng. Technol.* **2007**, *1*, 101–110. [CrossRef]
30. Naidu, V.P.S. Image Fusion Technique using Multi-resolution Singular Value Decomposition. *Def. Sci. J.* **2011**, *61*, 479. [CrossRef]
31. Hermessi, H.; Mourali, O.; Zagrouba, E. Multimodal medical image fusion review: Theoretical background and recent advances. *Signal Process.* **2021**, *183*, 108036. [CrossRef]
32. Wan, H.; Tang, X.; Zhu, Z.; Xiao, B.; Li, W. Multi-Focus Color Image Fusion Based on Quaternion Multi-Scale Singular Value Decomposition. *Front. Neurorobot.* **2021**, *15*, 76. [CrossRef]
33. Singh, S.; Anand, R.S. Multimodal Medical Image Sensor Fusion Model Using Sparse K-SVD Dictionary Learning in Nonsubsampled Shearlet Domain. *IEEE Trans. Instrum. Meas.* **2020**, *69*, 593–607. [CrossRef]

34. Liu, Y.; Chen, X.; Ward, R.K.; Wang, Z.J. Medical Image Fusion via Convolutional Sparsity Based Morphological Component Analysis. *IEEE Signal Process. Lett.* **2019**, *26*, 485–489. [CrossRef]
35. Maqsood, S.; Javed, U. Multi-modal Medical Image Fusion based on Two-scale Image Decomposition and Sparse Representation. *Biomed. Signal Process. Control* **2020**, *57*, 101810. [CrossRef]
36. Pankaj, D.; Sachin Kumar, S.; Mohan, N.; Soman, K.P. Image Fusion using Variational Mode Decomposition. *Indian J. Sci. Technol.* **2016**, *9*, 1–8. [CrossRef]
37. Vishnu Pradeep, V.; Sowmya, V.; Soman, K. Variational mode decomposition based multispectral and panchromatic image fusion. *IJCTA* **2016**, *9*, 8051–8059.
38. Pajares, G.; de la Cruz, J.M. A wavelet-based image fusion tutorial. *Pattern Recognit.* **2004**, *37*, 1855–1872. [CrossRef]
39. Ouahabi, A. *Signal and Image Multiresolution Analysis*; John Wiley & Sons: Hoboken, NJ, USA, 2012; ISBN 1118568664.
40. Nunes, J.; Bouaoune, Y.; Delechelle, E.; Niang, O.; Bunel, P. Image analysis by bidimensional empirical mode decomposition. *Image Vis. Comput.* **2003**, *21*, 1019–1026. [CrossRef]
41. Gilles, J. Empirical Wavelet Transform. IEEE Trans. *Signal Process.* **2013**, *61*, 3999–4010. [CrossRef]
42. Dragomiretskiy, K.; Zosso, D. Variational Mode Decomposition. *IEEE Trans. Signal Process.* **2013**, *62*, 531–544. [CrossRef]
43. Lahmiri, S.; Boukadoum, M. Biomedical image denoising using variational mode decomposition. In Proceedings of the 2014 IEEE Biomedical Circuits and Systems Conference (BioCAS), Lausanne, Switzerland, 22–24 October 2014; pp. 340–343. [CrossRef]
44. Lahmiri, S. Denoising techniques in adaptive multi-resolution domains with applications to biomedical images. *Health Technol. Lett.* **2017**, *4*, 25–29. [CrossRef]
45. Maheshwari, S.; Pachori, R.B.; Kanhangad, V.; Bhandary, S.V.; Acharya, U.R. Iterative variational mode decomposition based automated detection of glaucoma using fundus images. *Comput. Biol. Med.* **2017**, *88*, 142–149. [CrossRef]
46. Konstantin, D.; Zosso, D. Two-dimensional variational mode decomposition. In Proceedings of the International Workshop on Energy Minimization Methods in Computer Vision and Pattern Recognition, Hong Kong, China, 13–16 January 2015; Springer: Berlin/Heidelberg, Germany, 2015; pp. 197–208.
47. Polinati, S.; Dhuli, R. A review on multi-model medical image fusion. In Proceedings of the International Conference on Signal Processing, Communications and Computing (ICSPCC 2019), Liaoning, China, 20–22 September 2019; ICCSP: Tamilnadu, India, 2019.
48. Du, J.; Li, W.; Xiao, B. Anatomical-Functional Image Fusion by Information of Interest in Local Laplacian Filtering Domain. *IEEE Trans. Image Process.* **2017**, *26*, 5855–5866. [CrossRef]
49. Wang, Y.; Du, H.; Xu, J.; Liu, Y. A no-reference perceptual blur metric based on complex edge analysis. In Proceedings of the 2012 3rd IEEE International Conference on Network Infrastructure and Digital Content, Beijing, China, 21–23 September 2012; IEEE: New York, NY, USA, 2012; pp. 487–491.
50. Hossny, M.; Nahavandi, S.; Creighton, D. Comments on 'Information measure for performance of image fusion'. *Electron. Lett.* **2008**, *44*, 2–4. [CrossRef]
51. Sheikh, H.R.; Bovik, A.C. Image information and visual quality. *IEEE Trans. Image Process.* **2006**, *15*, 430–444. [CrossRef] [PubMed]
52. Ferroukhi, M.; Ouahabi, A.; Attari, M.; Habchi, Y.; Taleb-Ahmed, A. Medical Video Coding Based on 2nd-Generation Wavelets: Performance Evaluation. *Electronics* **2019**, *8*, 88. [CrossRef]
53. Xydeas, C.S.; Petrović, V. Objective image fusion performance measure. *Electron. Lett.* **2000**, *36*, 308. [CrossRef]
54. Wang, Z.; Bovik, A.C.; Sheikh, H.R.; Simoncelli, E.P. Image quality assessment: From error visibility to structural similarity. *IEEE Trans. Image Process.* **2004**, *13*, 600–612. [CrossRef]
55. Singh, R.; Khare, A. Multiscale medical image fusion in wavelet domain. *Sci. World J.* **2013**. [CrossRef]
56. Oliveira, F.P.M.; Tavares, J.M.R.S. Medical image registration: A review. *Comput. Methods Biomech. Biomed. Engin.* **2014**, *17*, 73–93. [CrossRef] [PubMed]

applied sciences

MDPI

Article

Consecutive Independence and Correlation Transform for Multimodal Data Fusion: Discovery of One-to-Many Associations in Structural and Functional Imaging Data

Chunying Jia [1,*], Mohammad Abu Baker Siddique Akhonda [1], Yuri Levin-Schwartz [2], Qunfang Long [1] and Vince D. Calhoun [3] and Tülay Adali [1]

1 Department of Computer Science and Electrical Engineering, University of Maryland Baltimore County, Baltimore, MD 21250, USA; mo32@umbc.edu (M.A.B.S.A.); qunfang1@umbc.edu (Q.L.); adali@umbc.edu (T.A.)
2 Icahn School of Medicine at Mount Sinai, New York, NY 10029, USA; yuri.levsch@gmail.com
3 Tri-Institutional Center for Translational Research in Neuroimaging and Data Science (TReNDS), Georgia State University, Georgia Institute of Technology, and Emory University, Atlanta, GA 30030, USA; vcalhoun@gsu.edu
* Correspondence: chunyin1@umbc.edu

Citation: Jia, C.; Akhonda, M.A.B.S.; Levin-Schwartz, Y.; Long, Q.; Calhoun, V.D.; Adali, T. Consecutive Independence and Correlation Transform for Multimodal Data Fusion: Discovery of One-to-Many Associations in Structural and Functional Imaging Data. *Appl. Sci.* **2021**, *11*, 8382. https://doi.org/10.3390/app11188382

Academic Editor: Michalis Vrigkas, Christophoros Nikou and Ioannis A. Kakadiaris

Received: 19 July 2021
Accepted: 4 September 2021
Published: 9 September 2021

Abstract: Brain signals can be measured using multiple imaging modalities, such as magnetic resonance imaging (MRI)-based techniques. Different modalities convey distinct yet complementary information; thus, their joint analyses can provide valuable insight into how the brain functions in both healthy and diseased conditions. Data-driven approaches have proven most useful for multimodal fusion as they minimize assumptions imposed on the data, and there are a number of methods that have been developed to uncover relationships across modalities. However, none of these methods, to the best of our knowledge, can discover "one-to-many associations", meaning one component from one modality is linked with more than one component from another modality. However, such "one-to-many associations" are likely to exist, since the same brain region can be involved in multiple neurological processes. Additionally, most existing data fusion methods require the signal subspace order to be identical for all modalities—a severe restriction for real-world data of different modalities. Here, we propose a new fusion technique—the consecutive independence and correlation transform (C-ICT) model—which successively performs independent component analysis and independent vector analysis and is uniquely flexible in terms of the number of datasets, signal subspace order, and the opportunity to find "one-to-many associations". We apply C-ICT to fuse diffusion MRI, structural MRI, and functional MRI datasets collected from healthy controls (HCs) and patients with schizophrenia (SZs). We identify six interpretable triplets of components, each of which consists of three associated components from the three modalities. Besides, components from these triplets that show significant group differences between the HCs and SZs are identified, which could be seen as putative biomarkers in schizophrenia.

Keywords: independent component analysis; independent vector analysis; multimodal data fusion; brain imaging

1. Introduction

Imaging and electrophysiological techniques that effectively quantify brain functions and structures facilitate our understanding of human brain function in healthy populations and those suffering from neuropsychiatric diseases such as schizophrenia [1–4]. For example, the magnetic resonance imaging (MRI)-based methods diffusion MRI (dMRI) [5], structural MRI (sMRI) [6], and functional MRI (fMRI) [7] are three brain-imaging modalities that convey information about structural connectivity, regional brain volume, and functional activity/connectivity, respectively. DMRI estimates white matter (WM) connectivity based on the differential diffusion rates of water molecules in different brain tissues. SMRI provides information about the morphology of brain tissues including WM, grey matter (GM), and cerebrospinal fluid (CSF). FMRI detects neuronal activations by measuring the blood oxygenation level-dependent response in the brain. Each of these three imaging techniques reveals different yet complementary information about brain structures or activities. Because of the intrinsic relationships between the structure and activity of the brain, multiple associations between these modalities are expected. This motivates the increasing popularity of collecting data from the same subjects using multiple modalities [8,9]. A joint analysis of such multimodal data should be useful for finding multiple associations across modalities, revealing the mechanisms of brain functions and potentially improving the predictive power to diagnose diseases compared with unimodal analyses. Data fusion is concerned with the simultaneous analyses of such joint datasets to obtain a global view of a problem under study or a system under observation by leveraging the information that is available across multiple datasets so that further analyses (e.g., detection, classification) can be improved [10–12].

However, since the relationships between different modalities are largely unknown, data-driven methods are particularly attractive. In this regard, data-driven methods based on blind source separation (BSS) have proven to be particularly useful for data fusion due to their ability to decompose the observed data into a set of latent variables, also known as components, through a simple generative model [13–16]. The components obtained from different datasets separately or jointly through the BSS techniques could be used to reveal potential relationships between different datasets. For example, components obtained from the datasets of sMRI and fMRI through the BSS techniques show locations and intensities of separated source signals in the brain (i.e., spatial maps) and allow us to find how the brain activities (fMRI) are related to the brain structures (sMRI). Importantly, due to the existence of systematic noise and interference in the real-world data, performing data fusion in the signal subspace with proper numbers of components (i.e., orders of signal subspace) effectively reduces the noise and interference and thus enables a better generalization ability.

Among various BSS techniques, independent component analysis (ICA) [17] and canonical correlation analysis (CCA) [18,19], as well as their extensions to multiple datasets—independent vector analysis (IVA) [20,21] and multi-set CCA (mCCA) [22,23]—have proven to be especially useful for data fusion. With a linear mixing model and the assumption of statistical independence, ICA decomposes observations into maximally independent components (ICs). IVA generalizes ICA to multiple data sets and utilizes the statistical dependence across different data sets to achieve a powerful solution for data fusion [20,24]. CCA and mCCA, on the other hand, make use of second-order statistics (correlation), and it can be shown that IVA generalizes both [21].

A number of models based on ICA, IVA, CCA, and mCCA have been developed for data fusion, such as joint ICA (jICA) [25], mCCA + jICA [26], and transposed IVA (tIVA) [15,27]. JICA concatenates all the datasets together and then performs ICA on the joint dataset for data fusion. The jICA model assumes a common mixing matrix and order as well as equal contributions across all datasets—a set of strong constraints, especially for more than two modalities. For mCCA + jICA, the use of mCCA in the first stage relaxes the strong assumptions of identical mixing matrices; however, a common order for the different modalities is still needed. TIVA also requires a common order in different modalities and a significantly large number of subjects for enough statistical power, which in a large number of cases is not feasible in current experimental settings. Collectively, none of these methods are flexible in terms of dealing with different orders and finding "one-to-many associations" in real-world data. However, firstly, due to the disparate measurements in different modalities, the order of the signal subspace is very likely to be different between modalities [28,29]. Secondly, one component in one modality might be associated with multiple components in other modalities because of the intrinsic relationships between modalities. For instance, the same WM connects multiple regions of the GM; thus, "one-to-many associations" between dMRI and sMRI datasets are expected. Besides, the fusion of more than two modalities of brain imaging data is expected to further facilitate exploiting joint information beyond pair-wise associations; nonetheless, most data fusion methods have been implemented to find multiple associations between only two modalities [26]. Therefore, a new framework for multi-modal data fusion is needed that is fully flexible in terms of the number of datasets combined, allowing for different orders across modalities and the discovery of "one-to-many associations".

In this study, we propose a fully flexible data fusion framework, consecutive independence and correlation transform (C-ICT), to jointly analyze more than two datasets. By successively performing ICA and IVA, C-ICT exploits the strengths of ICA and IVA for the joint analysis of multimodal data. First, ICA is performed on individual datasets separately to obtain maximally independent components and corresponding subject profile matrices (first-level mixing matrices). Second, meaningful ICs and the corresponding subject profiles are selected for further analysis. Third, IVA with a multivariate Gaussian model (IVA-G) [30] is performed on the subject profile matrices of different datasets to obtain the source component vectors (SCVs) and the second-level mixing matrices. SCVs that show significant pair-wise correlations are chosen for further analysis. Finally, we trace back to the ICs in the first stage based on subject profiles with the highest contribution to the correlated SCVs and identify them as associated components across different modalities. Thus, C-ICT is fully flexible in terms of the number of datasets combined, the numbers of orders of the signal subspace for each dataset, and the discovery of "one-to-many associations", and builds on the related concept we introduced in [28] for only two datasets.

We apply this new method to the fusion of data from three brain-imaging modalities (dMRI, sMRI, and resting-state fMRI) to search for possible multiple associations across these datasets. The data were collected from 86 healthy controls (HCs) and 76 patients with schizophrenia (SZs). Due to the intrinsic differences of these three data modalities, the choice of different orders for the three datasets is critical, which is accommodated by C-ICT. Importantly, prior to the IVA stage, we implemented an additional step to remove artifact components, as these might have contaminated the inherent associations across the modalities—a step which we note to be critical to the success of the method. It is shown that C-ICT successfully discovers multiple associations among the three modalities, including "one-to-many associations". First, multiple associations successfully discovered in this study are consistent with existing structure–function networks. Second, among the three modalities in each triplet, dMRI and sMRI components show stronger associations than other modality pairs (dMRI and fMRI; sMRI and fMRI), which is consistent with the fact that the GM structure and WM structure are more intrinsically related than other pairs. These results demonstrate that our proposed C-ICT method is a flexible and powerful tool to find associative relationships across related data of different modalities.

2. Materials and Methods

2.1. Human Brain Data

2.1.1. Data Acquisition

The data used in this study were dMRI, sMRI, and resting-state fMRI data from the Center of Biomedical Research Excellence (COBRE), available from the Collaborative Informatics and Neuroimaging Suite data exchange repository [31,32] (coins.trendscenter.org/, accessed on 8 October 2018). All the three datasets include the same 86 HCs (average age: 36.6 ± 12.1 years) and 76 SZs (average age: 36.9 ± 13.5 years). All patients completed the Structured Clinical Interview for DSM-IV Axis I Disorders (SCID [33]) for diagnostic confirmation (consensus was reached by two research psychiatrists using the SCID-DSM-IV-TR, patient version) and evaluation for co-morbidities [34]. The battery of cognitive performance tests used was the MATRICS (Measurement and Treatment Research to Improve Cognition in Schizophrenia) Cognitive Battery, including seven cognitive domains: speed of processing, attention/vigilance, working memory, verbal learning, visual learning, reasoning, and problem solving, and social cognition. We performed two-sample t-tests on these cognitive scores from the HC and SZ subjects. Compared with healthy controls, SZs had a significantly lower cognitive performance on all domains, supporting the findings that patients diagnosed with schizophrenia have an abnormal cognitive function.

DMRI data were collected along the anterior commissure–posterior commissure (AC–PC) line throughout the whole brain with a field of view (FOV) of 256 × 256 mm^2, a 128 × 128 matrix, 72 slices with a slice thickness of 2 mm (2 mm isotropic resolution), a number of excitations (NEX) of 1, echo time (TE) of 84 ms, and repetition time (TR) of 9000 ms. A multiple-channel radio-frequency (RF) coil was used, with GeneRalized Autocalibrating Partial Parallel Acquisition (GRAPPA) (X2), with 30 gradient directions with a b of 800 s/mm^2. The b = 0 experiment was repeated five times [35] and equally interspersed between the 30 gradient directions. The total scan time was about 6 min. This procedure was repeated twice to increase the signal-to-noise ratio (SNR).

For the sMRI data collection, the echo planar imaging (EPI) slices were collected in a sequential ascending order on a Siemens 3 T TIM Trio scanner using a 12-channel head coil. A sagittal gradient echo scout image through the mid-line was obtained to prescribe oblique axial image slices that were parallel to the AC–PC line. To minimize the susceptibility artifact in the orbitofrontal area, oblique slices were used [36]. High-resolution T1-weighted images were acquired with a five-echo multi-echo magnetization-prepared rapid gradient-echo (MP-RAGE) sequence (TE = 1.64, 3.5, 5.36, 7.22, 9.08 ms, TR = 2.53 s, TI (inversion time) = 1.2 s, flip angle = 7°, NEX = 1, slice thickness = 1 mm, FOV = 256 mm, resolution = 256 × 256) for region of interest analyses and spatial normalization.

The resting-state fMRI data were acquired using a conventional single-shot, gradient echo echo-planar pulse sequence with lipid suppression (TE = 29 ms; TR = 2000 ms; flip angle = 75°; FOV = 240 mm; matrix size = 64 × 64; 33 slices; voxel size: 3.75 × 3.75 × 4.55 mm^3). The first image of each run was removed to eliminate T1 equilibrium effects [37].The participants were instructed to keep their eyes open during the scan and stare passively at a central fixation cross for about 5 min.

2.1.2. Data Preprocessing and Feature Extraction

Neuroimaging data are usually high-dimensional, with each modality having different properties and dimensions. For example, sMRI data contain spatial information with a high spatial resolution but contain no temporal information. FMRI data convey both temporal information and spatial information, but with a lower spatial resolution. To jointly analyze different datasets in a common subspace, it is important to reduce data of each modality down to a single feature for each subject, a multivariate lower-dimensional representation of the data, and collect the features from all subjects together in a single dataset [13,15]. Such a reduction enables the datasets from all modalities to share a common dimension—i.e., the number of subjects—which provides a natural connection across different data types.

The dMRI data we used were preprocessed using the FMRIB Software Library (https://fsl.fmrib.ox.ac.uk/fsl/fslwiki/, accessed on 8 October 2018). We carried out the preprocessing steps as follows: (1) we identified and removed excessive motion or vibration artifacts through a quality check with any gradient directions; (2) we corrected motion and eddy currents; (3) we corrected gradient directions for any image rotation caused by the previous motion correction step; and (4) we calculated the diffusion tensor and then derived the scalar measure—fractional anisotropy (FA)—from the tensor. The FA, a measurement of WM integrity, was calculated as the fraction of total diffusion that can be attributed to anisotropic diffusion. A higher value of FA corresponds to more consistent diffusion orientation and thus more integrity of WM [38]. The FA maps were smoothed with an 8 mm full-width half-maximum (FWHM) Gaussian filter and then used as the feature for dMRI data.

Using the unified segmentation method in Statistical Parametric Mapping (SPM8) (http://www.fil.ion.ucl.ac.uk/spm, accessed on 8 October 2018), the sMRI data were (1) normalized to the Montreal Neurological Institute (MNI) space; (2) resliced to 3 × 3 × 3 mm^3; and (3) segmented into GM, WM, and CSF images. The segmented GM images were then smoothed with an 8 mm FWHM Gaussian filter. We further detected subject outliers using the spatial Pearson correlation with a template image to ensure that all images of subjects were properly segmented [39]. The GM images were used as the feature for sMRI data.

The fMRI data were preprocessed using an automated analysis pipeline [40] carried out in SPM8. The pipeline consisted of (1) aligning all the images with the first image as the reference using INRIalign approach [41] to correct minor motions of the subject; (2) correcting for time differences between the slices using the middle slice as the reference; and (3) spatial normalization to MNI space, including reslicing to $3 \times 3 \times 3$ mm^3, resulting in $53 \times 63 \times 46$ voxels. We then regressed out six motion parameters, WM, and CSF signals to further denoise the data. Data were then spatially smoothed with an 8 mm FWHM Gaussian filter. We estimated the fractional amplitude of low-frequency fluctuation (fALFF) [42], which is an effective approach to represent brain activity with high sensitivity and specificity that has been used in a large array of studies [43–48]. The fALFF is calculated as the ratio of the power spectrum of low frequency (0.0–0.08 Hz) to that of the whole detectable frequency range (0–0.25 Hz) [49]. The fALFF maps were then used as the feature for fMRI in this study.

For each modality, based on its own extracted feature, a two-dimensional matrix—feature dataset—was formed by concatenating the features across all subjects. These feature datasets from different modalities have a common dimension—the number of subjects—thus enabling the discovery of associations among different modalities through the variations across subjects.

2.2. *Background*

2.2.1. ICA

ICA is a statistical method that seeks to recover latent sources from a set of observed data with the assumption that the latent sources are statistically independent of one another [50]. Since it places few assumptions on data, ICA has been widely used in brain imaging studies [51–54].

Given a feature dataset $\mathbf{X} \in \mathbb{R}^{M \times V}$ comprised of M subjects and V samples (e.g., voxels), the generative model for noiseless ICA can be written as

$$\mathbf{X} = \mathbf{A}\mathbf{S}, \tag{1}$$

where $\mathbf{A} \in \mathbb{R}^{M \times M}$ is a full rank square mixing matrix and $\mathbf{S} \in \mathbb{R}^{M \times V}$ is the latent sources. The goal of ICA is to estimate a demixing matrix $\mathbf{W} \in \mathbb{R}^{M \times M}$ such that the estimated source matrix $\hat{\mathbf{S}}$ can be computed as

$$\hat{\mathbf{S}} = \mathbf{W}\mathbf{X}. \tag{2}$$

This can be achieved by minimizing the following cost function:

$$\mathcal{J}_{\text{ICA}}(\mathbf{W}) = \sum_{m=1}^{M} \mathcal{H}(\hat{\mathbf{s}}_m) - \log|\det(\mathbf{W})|, \tag{3}$$

where $\mathcal{H}(\cdot)$ is the differential entropy, $\hat{\mathbf{s}}_m$ denotes the mth row of $\hat{\mathbf{S}}$, and $\det(\mathbf{W})$ is the determinant of $\mathbf{W}$ [21]. Since both $\mathbf{A}$ and $\mathbf{S}$ are unknown, the solution of ICA decomposition is subject to permutation and scaling ambiguities. The scaling ambiguity can be resolved by normalizing the estimated ICs to have unit variance, as this is not information we can retain. Hence, the inverse of $\mathbf{W}$ can be considered to be an estimate of mixing matrix $\mathbf{A}$, subject to permutation ambiguity, which is fortunately not a serious problem in most applications. The columns of the estimated mixing matrix $\hat{\mathbf{A}}$—i.e., the inverse of $\mathbf{W}$—contain the weights for the estimated sources across subjects. We refer to these columns as the subject covariations or profiles and they can be used to explore the associations between different modalities.

2.2.2. IVA

IVA extends ICA to multiple datasets by exploiting the additional information from the statistical dependence across multiple datasets [21]. Given K datasets, each containing M observations and V samples, $\mathbf{X}^{[k]}$, the generative model for IVA assumes that each dataset is a linear mixture of M independent sources,

$$\mathbf{X}^{[k]} = \mathbf{A}^{[k]}\mathbf{S}^{[k]},\ 1 \leq k \leq K, \tag{4}$$

where $\mathbf{A}^{[k]} \in \mathbb{R}^{M\times M}$ denotes the kth mixing matrix and $\mathbf{S}^{[k]} \in \mathbb{R}^{M\times V}$ denotes the set of independent sources. IVA jointly estimates K demixing matrices, $\mathbf{W}^{[k]}$, to compute the estimated sources of each dataset,

$$\hat{\mathbf{S}}^{[k]} = \mathbf{W}^{[k]}\mathbf{X}^{[k]},\ 1 \leq k \leq K, \tag{5}$$

by minimizing the cost function given as

$$\mathcal{J}_{\mathrm{IVA}}(\mathbf{W}) = \sum_{m=1}^{M}\left[\sum_{k=1}^{K}\mathcal{H}\left(\hat{s}_m^{[k]}\right) - \mathcal{I}(\hat{\mathbf{s}}_m)\right] - \sum_{k=1}^{K}\log\left|\det \mathbf{W}^{[k]}\right|, \tag{6}$$

where $\mathcal{H}\left(\hat{s}_m^{[k]}\right)$ denotes the entropy of the mth source estimate for the kth dataset [21]. $\mathcal{I}(\hat{\mathbf{s}}_m)$ denotes the mutual information within the mth source component vector (SCV), defined as $\hat{\mathbf{s}}_m = [\hat{s}_m^{[1]}, \hat{s}_m^{[2]}, \ldots, \hat{s}_m^{[K]}]^{\mathrm{T}} \in \mathbb{R}^{K\times V}$, where $\hat{s}_m^{[k]} \in \mathbb{R}^{V}$ is the mth source from the kth dataset and the mth SCV is formed by concatenating the mth component from all the K datasets [24]. Thus, the maximization of the mutual information within the SCV enables IVA to exploit dependence across datasets. The estimated mixing matrix for the kth dataset is calculated as $\hat{\mathbf{A}}^{[k]} = (\mathbf{W}^{[k]})^{-1}$. The cth column of $\hat{\mathbf{A}}^{[k]}$ contains the weights for the cth row of the $\hat{\mathbf{S}}^{[k]}$ (cth source) across different rows of $\mathbf{X}^{[k]}$. Since the cth source is a part of the cth SCV, the cth column of $\hat{\mathbf{A}}^{[k]}$ can be used to identify which observation makes the highest contribution to the cth SCV; i.e., which observation makes the highest contribution to the dependence across different modalities. These identified observations from K modalities enable the discovery of associations across different modalities.

2.3. *C-ICT Framework*

As a hybrid model based on ICA and IVA-G, C-ICT factors and fuses multimodal data. By assuming that each underlying SCV has a multivariate Gaussian distribution, IVA-G takes into account only second-order statistics and thus maximizes correlation across different datasets. Due to its strong identifiability condition, IVA-G has been demonstrated to have superior performance to mCCA at identifying the correlation structure [21,24,30,55]. C-ICT exploits the advantages of ICA and IVA-G by independently decomposing each dataset and then fusing the subject covariations together to identify the associations among the modalities. Let $\mathbf{X}^{[k]} \in \mathbb{R}^{M\times V_k}$, $k = 1, 2, \ldots, K$ be feature datasets from K modalities, where M is the number of subjects, common across datasets, V_k is the number of voxels from the kth dataset, and the mth row of each feature dataset $\mathbf{X}^{[k]}$ represents the feature of the *mth* subject. A generative framework of C-ICT is shown in Figure 1. Briefly, C-ICT consists of the following steps: (1) ICA is used to estimate ICs and corresponding subject covariations for each dataset; (2) artifact components and corresponding subject covariations are eliminated; (3) IVA-G is used to explore the associations between subject covariations between modalities; and (4) associated subject covariations are used to identify associated ICs.

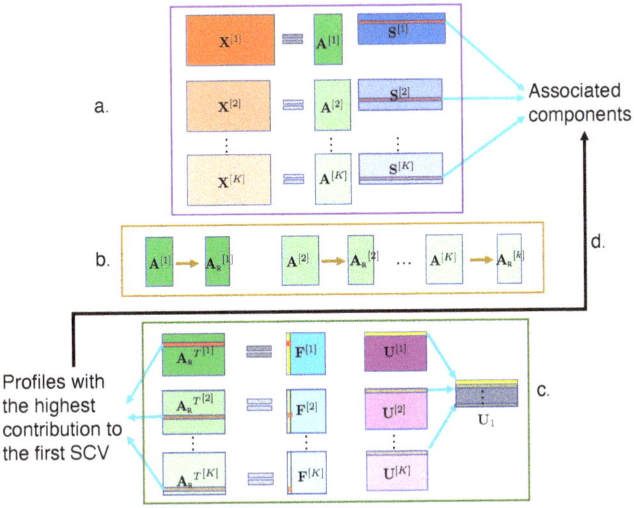

Figure 1. Flowchart of C-ICT. (**a**) Perform ICA on each dataset separately to obtain subject covariation matrices and ICs. (**b**) Eliminate subject covariations corresponding to artifact components. (**c**) Apply IVA-G on the reduced subject covariation matrices to obtain SCVs and the second-level mixing matrices. SCVs that show significant pair-wise correlation are chosen for further analysis. (**d**) Subject covariations with the highest contribution to the correlated SCVs chosen above are identified and their corresponding ICs are identified as the associated components across K modalities. The first SCV is highlighted for clarity. The same color of the bars in the different matrices denotes the same index in the corresponding rows and columns.

2.3.1. C-ICT Step 1: ICA

In the first stage, ICA is separately performed on each feature dataset to obtain maximally independent estimated source estimates, $\mathbf{S}^{[k]}$, and corresponding subject covariation matrices $\mathbf{A}^{[k]}, k = 1, 2, \ldots, K$, as shown in Figure 1a. The original ICA model assumes that the number of observations (subjects) is the same as the number of underlying sources. In practice, usually the number of observations is greater than the number of latent sources. Therefore, directly performing ICA on the original feature matrix might result in overfitting due to the effects of additive noise. So, it is desirable to reduce the dimensionality of the data to a lower-dimensional signal subspace and then perform ICA on the extracted signal subspace. Principal component analysis (PCA), a popular dimensionality reduction technique, is used to represent most of the variability across subjects. Let N_k be the order of the signal subspace—i.e., the number of estimated components—for the kth dataset. We first perform PCA on $\mathbf{X}^{[k]}$ to obtain dimension reduced matrices, $\mathbf{Y}^{[k]} \in \mathbb{R}^{N_k \times V_k}$, using the following:

$$\mathbf{Y}^{[k]} = \mathbf{V}^{T[k]} \mathbf{X}^{[k]}, \tag{7}$$

where $\mathbf{V}^{T[k]} \in \mathbb{R}^{N_k \times M}$ is the dimension reduction matrix. The ICA model applied to $\mathbf{Y}^{[k]}$ can be written as

$$\mathbf{Y}^{[k]} = \mathbf{A}_*{}^{[k]} \mathbf{S}^{[k]}, \tag{8}$$

where $\mathbf{A}_*{}^{[k]} \in \mathbb{R}^{N_k \times N_k}$ denotes the mixing matrix in the ICA model that is in the dimension reduced space. To obtain the back-reconstructed mixing matrix $\mathbf{A}^{[k]} \in \mathbb{R}^{M \times N_k}$ for the original mixture $\mathbf{X}^{[k]}$, we combine Equations (4), (7) and (8) to obtain the following equation:

$$\mathbf{A}_*{}^{[k]} = \left(\mathbf{V}^{[k]}\right)^T \mathbf{A}^{[k]}, \tag{9}$$

which is equivalent to

$$\mathbf{A}^{[k]} = \left(\mathbf{V}^{[k]}\right)^{+T} \mathbf{A}_*^{[k]}. \tag{10}$$

where $(\cdot)^{+T}$ denotes the transpose of the Moore–Penrose pseudo-inverse of a matrix. Note that this C-ICT framework allows for different orders for each dataset—an important feature for multimodal data fusion—as shown in Figure 2.

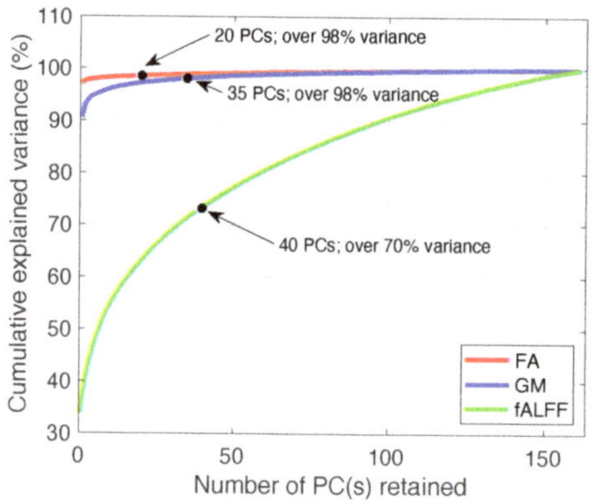

Figure 2. Cumulative explained variance accounting for different PCs from the FA, GM, and fALFF datasets. For each dataset, the black point on the curve indicates the variance explained by the number of PCs we selected.

Applying ICA to each dataset, k, results in a unique source matrix $\mathbf{S}^{[k]} \in \mathbb{R}^{N_k \times V_k}$ and estimated mixing matrix $\mathbf{A}^{[k]} \in \mathbb{R}^{M \times N_k}$ for each modality. Note again that the column of $\mathbf{A}^{[k]}$, denoted by $\mathbf{a}_j^{[k]}$, $j = 1, 2, \ldots, N_k$, represents the weights of the corresponding source $\mathbf{S}_j^{[k]}$ for each subject and enables the determination of possible connections between different modalities.

2.3.2. C-ICT Step 2: Artifact Elimination

We incorporated the "Artifact Elimination" step as part of the pipeline for C-ICT. In the second stage of C-ICT, artifact components are identified and eliminated, as shown in Figure 1b. This is feasible because physiological signals and artifact-related signals have independent causes and ICA has the capability to separate these statistically independent components; i.e., source signals of different origins. ICA has already been shown to be successful in separating real physiological source signals and artifact source signals such as those caused by motion [54,56,57]. In this study, we take advantage of this feature of ICA to eliminate the artifact components and avoid the contamination of the inherent associations between different modalities. In doing so, we obtain the reduced first-level mixing matrices $\mathbf{A}_R^{[k]} \in \mathbb{R}^{M \times L_k}, L_k \leq N_k$.

2.3.3. C-ICT Step 3: IVA

In the third stage, the second-level mixing matrices and SCVs, each of which comprises K second-level components that correspond to K modalities, are estimated by performing IVA-G on the transposed reduced first-level mixing matrices, as shown in Figure 1c. Let $\mathbf{B}^{[k]} = (\mathbf{A}_R^{[k]})^T \in \mathbb{R}^{L_k \times M}$ be $\mathbf{X}^{[k]}$ to be decomposed in the IVA generative model (4). The order of the subspace D is chosen to be $min(L_k)$. Let $\mathbf{F}^{[1]}, \mathbf{F}^{[2]}, \ldots, \mathbf{F}^{[K]}, \in \mathbb{R}^{L_k \times D}$ denote the estimated second-level mixing matrices and $\mathbf{U}^{[1]}, \mathbf{U}^{[2]}, \ldots, \mathbf{U}^{[K]}$ denote the estimated independent source matrices. The estimated sources are formed into SCVs $\mathbf{U}_1, \mathbf{U}_2, \ldots, \mathbf{U}_D$, and each $\in \mathbb{R}^{K \times M}$.

Among the estimated SCVs, we calculate all pairwise Pearson correlation coefficients ρ within each SCV. We then calculate the corresponding p-values; i.e., the probability (p_ρ) of obtaining a correlation as large as the observed value when the true correlation is zero. The correlation value ρ is considered significant if the p value is less than 0.05. If the K components within the jth SCV have significant pair-wise correlation coefficients (i.e., the p-values are all less than 0.05), then this jth SCV is referred to as a "significantly correlated SCV". All significantly correlated SCVs are used in the next steps. The number of these SCVs is denoted by C.

2.3.4. C-ICT Step 4: Tracing Back to Components

In the last stage, for each significantly correlated SCV, we identify K subject profiles with the highest contribution to that SCV from the K modalities, as shown in Figure 1c, and trace back to their corresponding ICs, as shown in Figure 1d. These ICs are defined as components that are associated across K modalities. The kth second-level source component in the cth SCV represents a linear combination of all rows of $\mathbf{B}^{[k]}$; i.e., a linear combination of all columns of $\mathbf{F}^{[k]}$, where $c = 1, 2, \ldots, C$. In other words, the kth second-level component in the cth SCV is a linear combination of all subject covariations from the kth modalities, weighted by the coefficients in the cth column of the second-level mixing matrix $\mathbf{F}^{[k]}$. The cth column of the estimated second-level mixing matrices, $\mathbf{F}^{[k]}$, represents the weights of the subject covariations for the cth SCV. We identify the indices $i^{[k]}$ of the largest absolute value of the coefficient in the cth column of $\mathbf{F}^{[k]}$ for the kth modality, respectively, implying the $i^{[k]}$th row of $\mathbf{F}^{[k]}$ has the highest contribution to the cth SCV. Then, the $i^{[1]}, i^{[2]}, \ldots, i^{[K]}$th ICs that correspond to the $i^{[k]}$th subject covariations obtained above are identified to be the cth associated ICs across K modalities.

Note that the back-reconstructed subject profiles are no longer necessarily orthogonal to each other from the same modality, so one subject profile might be dependent on multiple subject profiles from another modality. This enables C-ICT to be able to discover such "one-to-many associations" in this context; i.e., one IC from one modality may be associated with multiple ICs from another modality, which is a great advantage for the multimodal fusion method. The identification of "one-to-many associations" is illustrated in Figure 3. After identifying the associated subject profiles among K modalities, the ICs corresponding to these subject profiles are identified as associated among the K modalities; thus, the "one-to-many associations" are identified.

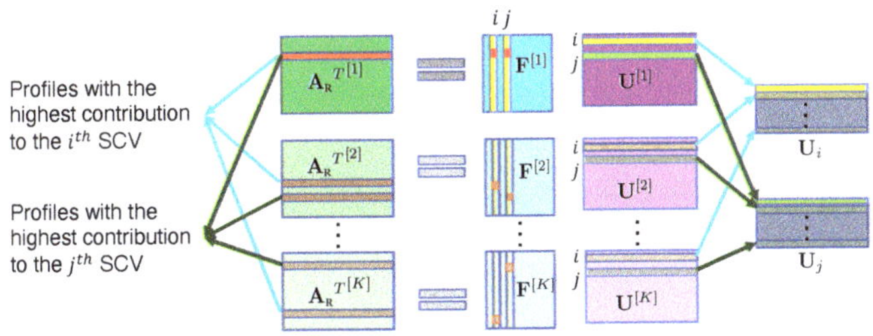

Figure 3. Identification of "one-to-many associations". In the first modality, the indices of the largest absolute value of the coefficient in the ith column and the jth column of $\mathbf{F}^{[1]}$ are the same, meaning that the same subject profile (marked in red) makes the highest contribution to the ith SCV and the jth SCV. This subject profile from the first modality is associated with the two subject profiles from the second modality and two subject profiles from the Kth modality.

3. Implementation and Results

3.1. Implementation

To exclude non-brain voxels, we performed masking on the FA, GM, and fALFF data separately using the Group ICA of fMRI Toolbox (GIFT) [58]. A matrix of the FA dataset with dimensions of 162 × 49,277 (number of subjects × number of dMRI voxels) was constructed for the dMRI dataset. A matrix of the GM dataset with dimensions of 162 × 60,636 (number of subjects × number of sMRI voxels) was constructed for the sMRI dataset. A matrix of the fALFF dataset with dimensions of 162 × 69,519 (number of subjects × number of fMRI voxels) was constructed for the fMRI dataset. Then, we implemented the proposed framework C-ICT to fuse the three datasets to explore the multiple associations across the dMRI, sMRI, and fMRI modalities.

3.1.1. Order Selection

We performed PCA on each dataset separately and calculated the cumulative explained variance (shown in Figure 2). From Figure 2, we can see that the variance retained for the FA and GM datasets follows a similar pattern, which differs from that of the fALFF dataset. For this reason, the criteria for order selection were similar for the FA and GM datasets. We used 98% as the threshold for the variance retained for the FA and GM datasets in order to balance the retention of the most signal while minimizing the effects of noise. As a result, the orders used for FA and GM datasets were 20 and 35, respectively.

Using a similar threshold for the variance retained with the fALFF dataset would result in an order of 145. Model order that is greater than 100 has been found to decrease the stability of ICA [59]. Therefore, using a variance retained threshold is not an appropriate way to determine the order for the fALFF dataset as was done for the FA and GM datasets. Instead, we determined the order for the fALFF dataset by investigating the effect of different orders on the estimated ICs [12,60–62]. We performed ICA to estimate ICs starting at an order of 20 and increasing the order with increments of 5 until the order reached 70. These limits were selected to cover a large range of potential, yet still reasonable, values for the order. As we increased the order from 20 to 40, we observed increasing numbers of meaningful neuroanatomical and functional components. However, when increasing the order beyond 40, we found that components of interest became split into multiple spatial maps and the number of the noise components increased. For this reason, we selected an order of 40 for the fALFF dataset, resulting in over 70% retained variance.

3.1.2. Algorithm Choice

Infomax [63] is a widely used ICA algorithm, particularly in the field of biomedical imaging. It assumes that the underlying source component has a super-Gaussian distribution, which is a good model-match for the brain imaging signals [64]. Therefore, we used Infomax for the C-ICT model in this study.

Since ICA is an iterative algorithm, the optimization of ICA yields different solutions depending on the initialization. Therefore, we performed ICA for 30 independent runs with different random initializations and selected the most consistent run using a metric called cross inter-symbol interference (cross-ISI) [65]. Similarly, IVA-G is also an iterative algorithm, so we performed IVA-G 50 times independently and used cross-ISI to select the most consistent run.

3.1.3. Artifact Elimination

After the ICA step, we obtained 20, 35, and 40 ICs for the FA, GM, and fALFF datasets, respectively. For dMRI, we used the ICBM-DTI-81 white-matter labels atlas and JHU white-matter tractography atlas [66–68], provided in FSL (fsl.fmrib.ox.ac.uk/fsl/fslwiki/, accessed on 10 November 2019) to identify the WM tracts. For sMRI and fMRI, each IC was transformed into Z-scores to obtain the zero mean and unit variance, and the voxels with a Z-score greater than the threshold 2 ($|Z| \geq 2$) were converted from MNI coordinates to Talairach coordinates and entered into a database to assign anatomical and functional labels for the left and right hemispheres.

To remove artifacts caused by body movements such as breathing and heart beating, we eliminated ICs from the brain ventricles or areas where large blood vessels are located [69,70]. We also eliminated ICs that were spread across wide regions of the brain or at the edges of brain images to remove motion-related signals [57,71]. More specifically, for the FA, we removed the ICs with spotty, diffusely distributed patterns over most of the brain, which represented noise. For the sMRI, we removed the sinus susceptibility noise and WM components [72]. For the fMRI, we removed ICs that were located beyond the area of GM; e.g., ventricles, skull, and surrounding tissue, WM, frontal/sagittal sinus susceptibility noise, CSF, and motion artifacts with ring patterns [72–77]. As a result, 10, 27, and 17 ICs were retained for FA, GM, and fALFF datasets, respectively. Their corresponding subject covariations were formed into three reduced subject covariation matrices with sizes of 10×162 (FA), 17×162 (fALFF), and 27×162 (GM), which were then used in the IVA step. The spatial maps of the retained components were converted to Z-scores and thresholded at $|Z| \geq 2$ (Figures 4–6).

3.1.4. Group Differences

After identifying the associated ICs by performing the C-ICT on the three modalities, we performed a two-sample t-test on the corresponding subject covariations to identify ICs that showed a statistically significant difference between the two groups (HC vs. SZ; $p < 0.05$), which are referred to as biomarkers of schizophrenia. Associated IC triplets that showed significant group differences in all dMRI, sMRI and fMRI modalities are referred to as associated biomarkers of disease.

3.2. Fusion Results

Following the explained procedure, we discovered six IC linked triplets (Figure 7). The identified brain regions of each identified fALFF IC are summarized in Supplementary Table S1. Within each of the six triplets, three ICs that corresponded to three modalities (dMRI, sMRI and fMRI) were linked, representing a multimodal brain network with three nodes (WM, structural GM, and functional GM).

We found that the structure–structure associations were generally stronger than the structure–function associations (dMRI–sMRI: $|\rho| = 0.48 \pm 0.26$; dMRI–fMRI: $|\rho| = 0.32 \pm 0.11$; sMRI–fMRI: $|\rho| = 0.33 \pm 0.13$; see Table 1). We also examined group differences (HC vs. SZ) for each IC on its corresponding subject covariation and found that more ICs showed significant group differences in sMRI than the other two modalities (5 in sMRI; 2 in dMRI; 1 in fMRI; $p < 0.05$; two-sample t-test; see Figure 7). Interestingly, for all these ICs in both the dMRI and the sMRI data that showed significant group differences, the SZ group showed less activation, suggesting both a reduced volume of the GM and reduced integrity of the WM in patients with SZ. For the ICs in fMRI that showed significant group differences, the SZ group had higher activation, perhaps suggesting less neural efficiency in patients with SZ.

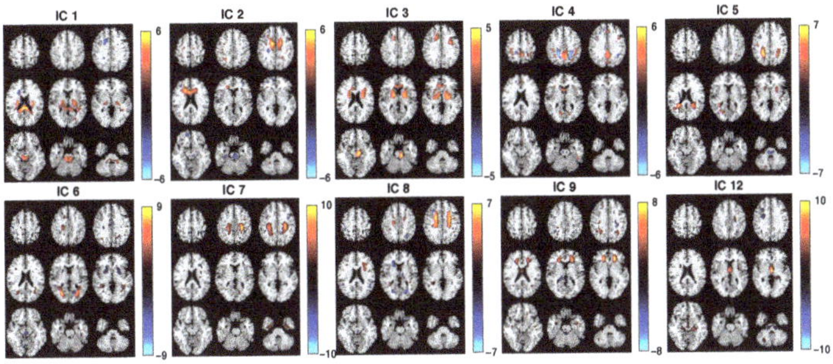

Figure 4. Spatial maps of 10 retained ICs from the FA dataset. These results reflect the IC results of SZ and HC groups combined. The brain maps are visualized at $|Z| \geq 2$, with positive Z values in red color and negative Z values in blue.

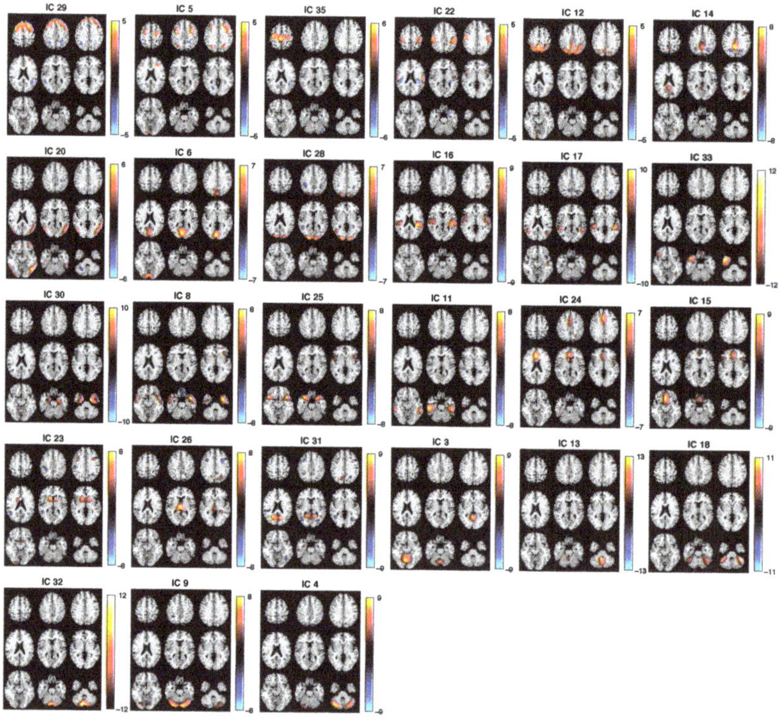

Figure 5. Spatial maps of 27 retained ICs from the GM dataset. These results reflect the IC results of SZ and HC groups combined. The brain maps are visualized at $|Z| \geq 2$, with positive Z values in red color and negative Z values in blue.

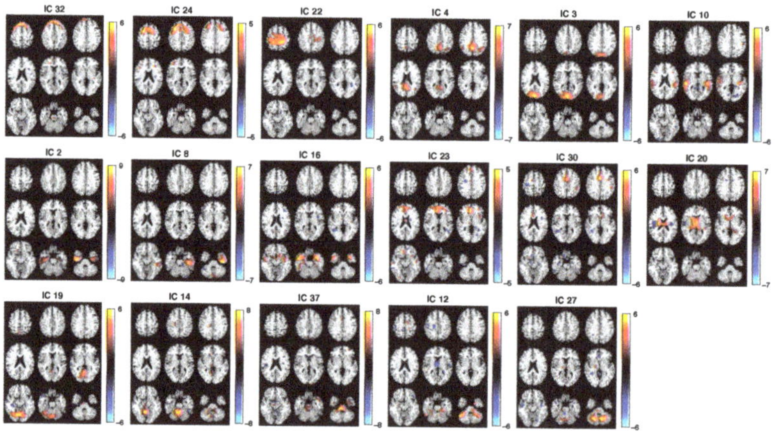

Figure 6. Spatial maps of 17 retained ICs from the fALFF dataset. These results reflect the IC results of SZ and HC groups combined. The brain maps are visualized at $|Z| \geq 2$, with positive Z values in red color and negative Z values in blue.

We also note that one IC can be found in more than one triplet, meaning that one IC from one modality is associated with many ICs from other modalities. For instance, the IC (corticospinal tract and superior longitudinal fasciculus (CST-SLF)) from dMRI is not only associated with the uncus and inferior temporal gyrus (Uncus–ITG) from sMRI and superior frontal gyrus and middle frontal gyrus (SFG-MFG) from fMRI, but is also associated with the precuneus and paracentral lobule (Precuneus–PCL) from sMRI and superior temporal gyrus (STG) from fMRI, as shown in Figure 7 (green frame). Another shared IC between triplets is the postcentral gyrus and precentral gyrus (POG–PCG) from sMRI, which is not only associated with anterior thalamic radiation and the anterior corona radiata (ATR–ACR) from dMRI and thalamus from fMRI, but also associated with superior corona radiation and the corticospinal tract (SCR–CST) from dMRI and culmen from fMRI, as shown in Figure 7 (blue frame).

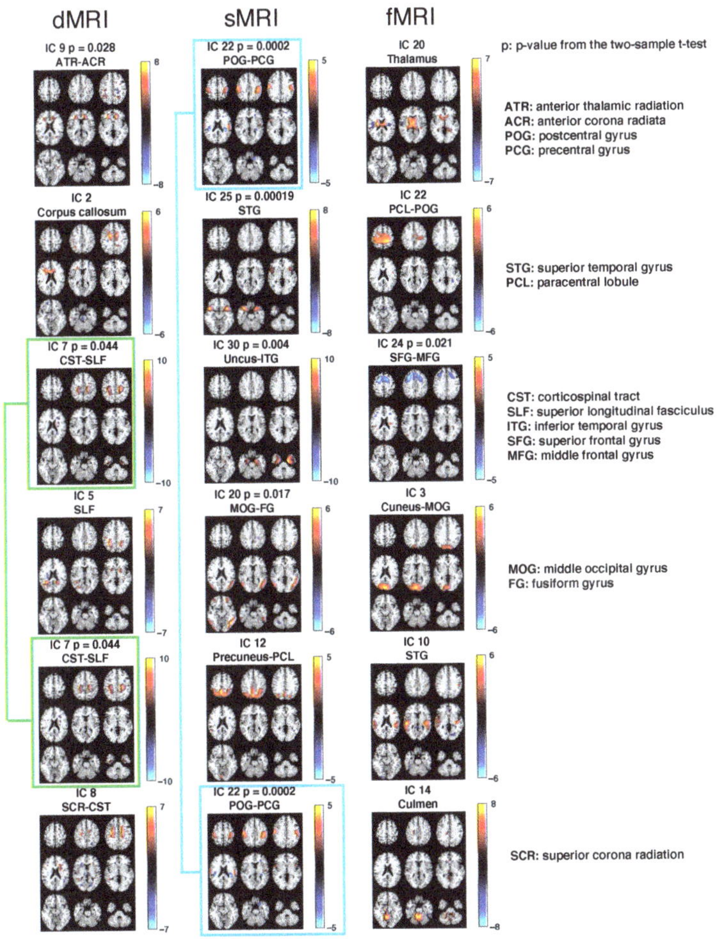

Figure 7. Summary of the six IC linked triplets. The spatial maps of all ICs were converted to Z-scores and thresholded by $|Z| \geq 2$. Each row shows the three associated ICs from dMRI, sMRI, and fMRI, respectively. The abbreviations of brain regions identified by ICs are shown above the spatial maps. If one IC shows a statistically significant activation between HCs and SZs ($p < 0.05$; t-test), the p value would be shown above the map, where the red color indicates higher activation in HCs than patients and the blue color indicates lower activation in HCs than patients. Note that the IC 7 (green frame) from dMRI is associated with two ICs from sMRI and two ICs from fMRI. IC 22 (blue frame) from sMRI is associated with two ICs from dMRI and two ICs from fMRI.

Table 1. The absolute values of the pair-wise Pearson correlation coefficients $|\rho|$ corresponding to the associated triplets and the values of p_ρ.

Triplet #	dMRI–sMRI	dMRI–fMRI	sMRI–fMRI
1	$\|\rho\| = 0.80, p_\rho = 4 \times 10^{-37}$	$\|\rho\| = 0.27, p_\rho = 5.6 \times 10^{-4}$	$\|\rho\| = 0.49, p_\rho = 3.6 \times 10^{-1}$
2	$\|\rho\| = 0.67, p_\rho = 4.2 \times 10^{-22}$	$\|\rho\| = 0.43, p_\rho = 1.2 \times 10^{-8}$	$\|\rho\| = 0.38, p_\rho = 1.9 \times 10^{-4}$
3	$\|\rho\| = 0.62, p_\rho = 1.8 \times 10^{-18}$	$\|\rho\| = 0.32, p_\rho = 3.8 \times 10^{-5}$	$\|\rho\| = 0.21, p_\rho = 4 \times 10^{-3}$
4	$\|\rho\| = 0.43, p_\rho = 1.6 \times 10^{-8}$	$\|\rho\| = 0.21, p_\rho = 7.7 \times 10^{-3}$	$\|\rho\| = 0.2, p_\rho = 1.2 \times 10^{-2}$
5	$\|\rho\| = 0.18, p_\rho = 2.6 \times 10^{-2}$	$\|\rho\| = 0.21, p_\rho = 6.9 \times 10^{-3}$	$\|\rho\| = 0.43, p_\rho = 8.4 \times 10^{-9}$
6	$\|\rho\| = 0.18, p_\rho = 2 \times 10^{-2}$	$\|\rho\| = 0.46, p_\rho = 1.1 \times 10^{-9}$	$\|\rho\| = 0.25, p_\rho = 1.2 \times 10^{-3}$

4. Discussion

In this study, we propose a new data fusion framework, C-ICT, to uncover relationships across multiple brain imaging modalities to reveal the underlying mechanisms of brain function as well as their dysfunction in diseases such as schizophrenia. Specifically, C-ICT uses ICA to extract independent brain networks from each modality and then explores possible associations across the modalities using IVA.

Compared with other existing data fusion frameworks, C-ICT is flexible in many ways. First, unlike jICA, mCCA + jICA, tIVA, and parallel ICA [15,25–27,78], C-ICT allows for different orders from the different datasets. Datasets of different modalities are different in nature and might also have different levels of signal-to-noise ratios (SNRs). Therefore, different orders would be expected for each modality. Second, C-ICT has no constraint on the number of datasets, like parallel ICA. This freedom is important because a greater number of data modalities can be collected from the same subjects. Thus, a framework that is capable of jointly analyzing all modalities at the same time can provide valuable insight into both the structural and functional aspects of networks responsible for brain functions. Third, C-ICT is unique in that it is able to discover "one-to-many associations" between datasets—a feature not available for any other data fusion method. This is possible for C-ICT because one IC with the highest contribution to one SCV might also have the highest contribution to other SCVs. The finding of such "one-to-many associations" suggests that one brain region can be recruited by multiple other regions for the performance of complex functions—an important neurobiological mechanism that is also supported by other studies [79–81]. In addition to being highly flexible, another advantage of the C-ICT framework is a critical step that eliminates ICs that might originate from artifacts such as heartbeats, respiratory movement, and non-brain signals. In doing so, we have observed more interpretable triplets of components, where each of the ICs from the three modalities are associated with each other, and stronger associations between modalities within each triplet, when compared with an implementation that skips this step.

We have applied the C-ICT framework to brain imaging data that measure structural and functional aspects of both the WM and GM through dMRI, sMRI, and fMRI in both HCs and SZs. As shown in the results, C-ICT uncovered six IC triplets. Importantly, within each triplet, the three associated components are observed to be all closely functionally related, thus validating our approach.

The first triplet includes anterior thalamic radiation and anterior corona radiata (ATR–ACR), postcentral gyrus and precentral gyrus (POG–PCG), and thalamus from dMRI, sMRI, and fMRI, respectively. The ATR is a WM bundle that connects the thalamus with the prefrontal cortex [82]. The thalamus is a hub that relays sensory information to the cerebral cortex [83]. The ACR carries somatotopically arranged motor fibers away from the cerebral cortex. The POG is where the primary somatosensory cortex is located, and PCG is where the primary motor cortex is located. This triplet represents a multimodal brain network that involves nodes that are all related to sensory–motor functions.

The second triplet includes the corpus callosum, superior temporal gyrus (STG), and paracentral lobule and postcentral gyrus (PCL–POG) from dMRI, sMRI, and fMRI, respectively. The corpus callosum is an important fiber bundle that connects the left and right brain hemispheres and allows communications of sensory, motor, and cognitive information between the two hemispheres. The STG is a cortical area responsible for auditory processing, multisensory integration, and spoken word recognition. The PCL is important for motor functions, while the POG is the location of the primary somatosensory cortex, which is responsible for the sense of touch. Thus, this triplet primarily focuses on sensory functions.

The third triplet includes the corticospinal tract and superior longitudinal fasciculus (CST–SLF), uncus and inferior temporal gyrus (Uncus–ITG), and superior frontal gyrus and middle frontal gyrus (SFG–MFG) from dMRI, sMRI, and fMRI, respectively. The CST is a fiber tract that originates from the motor-related cerebral cortex to the motor neurons and interneurons in the spinal cord and controls body movement. The SLF is a fiber tract that connects frontal, occipital, parietal, and temporal lobes including the premotor cortex, thus playing a role in regulating cognitive functions and motor behavior. The ITG is the anterior region of the temporal lobe, which is important for object cognition and memory. SFG and MFG occupy a large proportion of the frontal area that is also involved in cognition and motor functions. The three ICs within this triplet all show significant group differences between the HCs and SZs, suggesting that the structural and functional changes in these brain cognition and motor networks might be related to the cognitive dysfunction and motor deficits in SZs, thus revealing a multimodal neuroimaging biomarker of SZ.

The fourth triplet includes the superior longitudinal fasciculus (SLF), middle occipital gyrus and fusiform gyrus (MOG–FG), and cuneus and middle occipital gyrus (Cuneus–MOG) from dMRI, sMRI, and fMRI, respectively. The FG a large gyrus that spans across the temporal and occipital lobes. Both MOG and FG are involved in visual processing and cognition. Besides, the SLF is a fiber tract that connects to brain regions that are involved in cognitive function. Thus, this triplet might primarily process visual cognition such as object and facial recognition.

The fifth triplet includes CST–SLF, Precuneus–PCL, and STG from dMRI, sMRI, and fMRI, respectively. While both the structural components are involved in motor functions, the functional component is involved in sensory processing, representing a sensory–motor network across the three modalities. The sixth triplet includes SCR–CST, POG–PCG, and culmen from dMRI, sMRI, and fMRI, respectively. The three components are all related to motor function.

As we can see from the description of the six triplets, all triplets represent networks that are involved, at least partially, in motor functions, and many are related to cognitive functions. Importantly, we also find that the same IC appears in more than one triplet. Specifically, CST–SLF is in both triplet 3 (CST–SLF, Uncus–ITG, and SFG–MFG) and triplet 5 (CST–SLF, Precuneus–PCL, STG), and POG–PCG is in both triplet 1 (ATR–ACR, POG–PCG, and thalamus) and triplet 6 (SCR–CST, POG–PCG, and culmen), suggesting dual functions of one brain region in multiple multimodal brain networks.

Additionally, exploring the differences of ICs within the six triplets between HCs and SZs might provide information about potential biomarkers of schizophrenia. By using a two-sample *t*-test, we find that the ATR–ACR and CST–SLF from dMRI have stronger activation in HCs than in SZs, implying the WM in these regions is less integrated in SZs than in HCs. For sMRI, we find POG–PCG, STG, Uncus–ITG, and MOG–FG are less activated in SZs than in HCs, meaning that SZs have a reduced average GM volume in these areas. Furthermore, we find that SFG–MFG from fMRI has a stronger activation in SZs than HCs. Most of these regions has been reported to be associated with abnormalities and negative symptoms in SZs. SZs have been shown to have significantly lower FA values than HCs in the ATR [82,84–86], ACR [87], CST [88], and SLF regions [89]. In addition, the GM volume in the POG [90], PCG [91], STG [92], uncus [93], ITG [94,95], MOG [96], and FG [97] has been shown to be significantly decreased in SZs compared to in HCs.

5. Conclusions

It is increasingly common for multimodal data to be collected from the same subjects. This provides the motivation for the development of multimodal data fusion techniques, which have become important for the understanding of human brain imaging. In this paper, we proposed a novel multimodal data fusion framework, C-ICT, to explore multiple associations across different data modalities. We applied C-ICT to discover underlying relationships between dMRI, sMRI, and fMRI datasets from HCs and SZs. We have shown that C-ICT reveals multiple associations across the three modalities and provides potential biomarkers for schizophrenia, demonstrating that C-ICT is a flexible and informative method for the fusion of medical imaging data from different modalities. The success of C-ICT in this paper motivates its application to other modalities and domains. Besides the fusion of dMRI, sMRI, and fMRI data, C-ICT can be also used to fuse other types of multimodal data, such as magnetoencephalography (MEG) or genetic data. Moreover, this approach is not limited to the fusion of three data modalities but can be used for the fusion of many more modalities, thus enabling the discovery of more comprehensive associations across modalities in brain imaging studies as well as in other related areas.

Supplementary Materials: The following are available online at https://www.mdpi.com/article/10.3390/app11188382/s1, Table S1: Detailed information about the identified fALFF components.

Author Contributions: Conceptualization, C.J. and T.A.; Data curation, C.J. and V.D.C.; Formal analysis, C.J.; Funding acquisition, V.D.C. and T.A.; Investigation, C.J.; Methodology, C.J., M.A.B.S.A., Y.L.-S., Q.L. and T.A.; Project administration, T.A.; Resources, V.D.C. and T.A.; Software, C.J., M.A.B.S.A., Y.L.-S., Q.L. and T.A.; Supervision, T.A.; Validation, C.J.; Visualization, C.J. and T.A.; Writing—original draft, C.J.; Writing—review and editing, C.J., M.A.B.S.A., Y.L.-S., Q.L., V.D.C. and T.A. All authors have read and agreed to the published version of the manuscript.

Funding: This work was supported by the National Science Foundation (grants NSF-CCF 1618551, NSF-NCS 1631838) and National Institutes of Health grant R01-MH118695, R01-MH123610-01.

Institutional Review Board Statement: Not applicable.

Informed Consent Statement: Informed consent was obtained from all subjects involved in the study.

Data Availability Statement: The data used in this study are from the Center of Biomedical Research Excellence (COBRE), available from the Collaborative Informatics and Neuroimaging Suite data exchange repository [31,32] (coins.trendscenter.org/, accessed on 8 October 2018).

Acknowledgments: The authors would like to thank the research staff members from the Mind Research Network COBRE study for collecting, preprocessing, and sharing the data. The authors also appreciate the valuable feedback given by the members of Machine Learning for Signal Processing Laboratory at the University of Maryland, Baltimore County.

Conflicts of Interest: The authors declare no conflict of interest.

References

1. Le Bihan, D.; Johansen-Berg, H. Diffusion MRI at 25: Exploring brain tissue structure and function. *NeuroImage* **2012**, *61*, 324–341. [CrossRef]
2. Wang, Z.; Dai, Z.; Gong, G.; Zhou, C.; He, Y. Understanding structural-functional relationships in the human brain: A large-scale network perspective. *Neuroscientist* **2015**, *21*, 290–305. [CrossRef] [PubMed]
3. Birur, B.; Kraguljac, N.V.; Shelton, R.C.; Lahti, A.C. Brain structure, function, and neurochemistry in schizophrenia and bipolar disorder—A systematic review of the magnetic resonance neuroimaging literature. *NPJ Schizophr.* **2017**, *3*, 1–15. [CrossRef]
4. Dumoulin, S.O.; Fracasso, A.; van der Zwaag, W.; Siero, J.C.; Petridou, N. Ultra-high field MRI: Advancing systems neuroscience towards mesoscopic human brain function. *NeuroImage* **2018**, *168*, 345–357. [CrossRef]
5. Le Bihan, D.; Mangin, J.F.; Poupon, C.; Clark, C.A.; Pappata, S.; Molko, N.; Chabriat, H. Diffusion tensor imaging: Concepts and applications. *J. Magn. Reson. Imaging Off. J. Int. Soc. Magn. Reson. Med.* **2001**, *13*, 534–546. [CrossRef]
6. Coffman, J.A.; Bornstein, R.A.; Olson, S.C.; Schwarzkopf, S.B.; Nasrallah, H.A. Cognitive impairment and cerebral structure by MRI in bipolar disorder. *Biol. Psychiatry* **1990**, *27*, 1188–1196. [CrossRef]
7. Biswal, B.; Zerrin Yetkin, F.; Haughton, V.M.; Hyde, J.S. Functional connectivity in the motor cortex of resting human brain using echo-planar MRI. *Magn. Reson. Med.* **1995**, *34*, 537–541. [CrossRef] [PubMed]

8. Wang, S.; Fan, G. Alterations of structural and functional connectivity in profound sensorineural hearing loss infants within an early sensitive period: A combined DTI and fMRI study. *Dev. Cogn. Neurosci.* **2019**, *38*, 100654. [CrossRef]
9. Hirjak, D.; Rashidi, M.; Kubera, K.M.; Northoff, G.; Fritze, S.; Schmitgen, M.M.; Sambataro, F.; Calhoun, V.D.; Wolf, R.C. Multimodal magnetic resonance imaging data fusion reveals distinct patterns of abnormal brain structure and function in catatonia. *Schizophr. Bull.* **2020**, *46*, 202–210. [CrossRef]
10. Sui, J.; Yu, Q.; He, H.; Pearlson, G.; Calhoun, V.D. A selective review of multimodal fusion methods in schizophrenia. *Front. Hum. Neurosci.* **2012**, *6*, 27. [CrossRef] [PubMed]
11. Lahat, D.; Adali, T.; Jutten, C. Multimodal data fusion: An overview of methods, challenges, and prospects. *Proc. IEEE* **2015**, *103*, 1449–1477. [CrossRef]
12. Adali, T.; Levin-Schwartz, Y.; Calhoun, V.D. Multimodal data fusion using source separation: Application to medical imaging. *Proc. IEEE* **2015**, *103*, 1494–1506. [CrossRef]
13. Calhoun, V.D.; Adali, T. Feature-based fusion of medical imaging data. *IEEE Trans. Inf. Technol. Biomed.* **2008**, *13*, 711–720. [CrossRef]
14. James, A.P.; Dasarathy, B.V. Medical image fusion: A survey of the state of the art. *Inf. Fusion* **2014**, *19*, 4–19. [CrossRef]
15. Adali, T.; Levin-Schwartz, Y.; Calhoun, V.D. Multimodal data fusion using source separation: Two effective models based on ICA and IVA and their properties. *Proc. IEEE* **2015**, *103*, 1478–1493. [CrossRef]
16. Adali, T.; Akhonda, M.; Calhoun, V.D. ICA and IVA for data fusion: An overview and a new approach based on disjoint subspaces. *IEEE Sens. Lett.* **2018**, *3*, 1–4. [CrossRef] [PubMed]
17. Comon, P.; Jutten, C. *Handbook of Blind Source Separation: Independent Component Analysis and Applications*; Academic Press: Cambridge, MA, USA, 2010.
18. Harold, H. Relations between two sets of variates. *Biometrika* **1936**, *28*, 321–377.
19. Correa, N.M.; Li, Y.O.; Adali, T.; Calhoun, V.D. Canonical correlation analysis for feature-based fusion of biomedical imaging modalities and its application to detection of associative networks in schizophrenia. *IEEE J. Sel. Top. Signal Process.* **2008**, *2*, 998–1007. [CrossRef] [PubMed]
20. Kim, T.; Eltoft, T.; Lee, T.W. Independent vector analysis: An extension of ICA to multivariate components. In Proceedings of the International Conference on Independent Component Analysis and Signal Separation, Charleston, SC, USA, 5–8 March 2006; Springer: Berlin/Heidelberg, Germany, 2006; pp. 165–172.
21. Adali, T.; Anderson, M.; Fu, G.S. Diversity in independent component and vector analyses: Identifiability, algorithms, and applications in medical imaging. *IEEE Signal Process. Mag.* **2014**, *31*, 18–33. [CrossRef]
22. Kettenring, J.R. Canonical analysis of several sets of variables. *Biometrika* **1971**, *58*, 433–451. [CrossRef]
23. Correa, N.M.; Eichele, T.; Adali, T.; Li, Y.; Calhoun, V.D. Multi-set canonical correlation analysis for the fusion of concurrent single trial ERP and functional MRI. *NeuroImage* **2010**, *50*, 1438–1445. [CrossRef]
24. Anderson, M.; Fu, G.S.; Phlypo, R.; Adali, T. Independent vector analysis: Identification conditions and performance bounds. *IEEE Trans. Signal Process.* **2014**, *62*, 4399–4410. [CrossRef]
25. Calhoun, V.D.; Adali, T.; Pearlson, G.; Kiehl, K.A. Neuronal chronometry of target detection: Fusion of hemodynamic and event-related potential data. *NeuroImage* **2006**, *30*, 544–553. [CrossRef] [PubMed]
26. Sui, J.; He, H.; Pearlson, G.D.; Adali, T.; Kiehl, K.A.; Yu, Q.; Clark, V.P.; Castro, E.; White, T.; Mueller, B.A.; et al. Three-way (N-way) fusion of brain imaging data based on mCCA+ jICA and its application to discriminating schizophrenia. *NeuroImage* **2013**, *66*, 119–132. [CrossRef] [PubMed]
27. Levin-Schwartz, Y.; Calhoun, V.D.; Adali, T. Data-driven fusion of EEG, functional and structural MRI: A comparison of two models. In Proceedings of the 2014 48th Annual Conference on Information Sciences and Systems (CISS), Princeton, NJ, USA, 19–21 March 2014; pp. 1–6.
28. Akhonda, M.A.B.S.; Levin-Schwartz, Y.; Bhinge, S.; Calhoun, V.D.; Adali, T. Consecutive independence and correlation transform for multimodal fusion: Application to EEG and fMRI data. In Proceedings of the 2018 IEEE International Conference on Acoustics, Speech and Signal Processing (ICASSP), Calgary, AB, Canada, 15–20 April 2018; pp. 2311–2315.
29. Jia, C.; Akhonda, M.A.B.S.; Long, Q.; Calhoun, V.D.; Waldstein, S.; Adali, T. C-ICT for Discovery of Multiple Associations in Multimodal Imaging Data: Application to Fusion of fMRI and DTI Data. In Proceedings of the 2019 53rd Annual Conference on Information Sciences and Systems (CISS), Baltimore, MD, USA, 20–22 March 2019; pp. 1–5.
30. Anderson, M.; Adali, T.; Li, X.L. Joint blind source separation with multivariate Gaussian model: Algorithms and performance analysis. *IEEE Trans. Signal Process.* **2011**, *60*, 1672–1683. [CrossRef]
31. Scott, A.; Courtney, W.; Wood, D.; De la Garza, R.; Lane, S.; Wang, R.; King, M.; Roberts, J.; Turner, J.A.; Calhoun, V.D. COINS: An innovative informatics and neuroimaging tool suite built for large heterogeneous datasets. *Front. Neuroinform.* **2011**, *5*, 33. [CrossRef]
32. Wood, D.; King, M.; Landis, D.; Courtney, W.; Wang, R.; Kelly, R.; Turner, J.A.; Calhoun, V.D. Harnessing modern web application technology to create intuitive and efficient data visualization and sharing tools. *Front. Neuroinform.* **2014**, *8*, 71. [CrossRef] [PubMed]
33. Bell, C.C. DSM-IV: Diagnostic and statistical manual of mental disorders. *JAMA* **1994**, *272*, 828–829. [CrossRef]

34. Aine, C.; Bockholt, H.J.; Bustillo, J.R.; Cañive, J.M.; Caprihan, A.; Gasparovic, C.; Hanlon, F.M.; Houck, J.M.; Jung, R.E.; Lauriello, J.; et al. Multimodal neuroimaging in schizophrenia: Description and dissemination. *Neuroinformatics* **2017**, *15*, 343–364. [CrossRef]
35. Jones, D.K.; Horsfield, M.A.; Simmons, A. Optimal strategies for measuring diffusion in anisotropic systems by magnetic resonance imaging. *Magn. Reson. Med. Off. J. Int. Soc. Magn. Reson. Med.* **1999**, *42*, 515–525. [CrossRef]
36. Deichmann, R.; Gottfried, J.A.; Hutton, C.; Turner, R. Optimized EPI for fMRI studies of the orbitofrontal cortex. *NeuroImage* **2003**, *19*, 430–441. [CrossRef]
37. Shin, J.; Ahn, S.; Hu, X. Correction for the T1 effect incorporating flip angle estimated by Kalman filter in cardiac-gated functional MRI. *Magn. Reson. Med.* **2013**, *70*, 1626–1633. [CrossRef]
38. Assaf, Y.; Pasternak, O. Diffusion tensor imaging (DTI)-based white matter mapping in brain research: A review. *J. Mol. Neurosci.* **2008**, *34*, 51–61. [CrossRef]
39. Van Erp, T.G.; Greve, D.N.; Rasmussen, J.; Turner, J.; Calhoun, V.D.; Young, S.; Mueller, B.; Brown, G.G.; McCarthy, G.; Glover, G.H.; et al. A multi-scanner study of subcortical brain volume abnormalities in schizophrenia. *Psychiatry Res. Neuroimaging* **2014**, *222*, 10–16. [CrossRef]
40. Bockholt, H.J.; Scully, M.; Courtney, W.; Rachakonda, S.; Scott, A.; Caprihan, A.; Fries, J.; Kalyanam, R.; Segall, J.; De La Garza, R.; et al. Mining the mind research network: A novel framework for exploring large scale, heterogeneous translational neuroscience research data sources. *Front. Neuroinform.* **2010**, *3*, 36. [CrossRef] [PubMed]
41. Freire, L.; Roche, A.; Mangin, J.F. What is the best similarity measure for motion correction in fMRI time series? *IEEE Trans. Med. Imaging* **2002**, *21*, 470–484. [CrossRef] [PubMed]
42. Zou, Q.; Zhu, C.; Yang, Y.; Zuo, X.; Long, X.; Cao, Q.; Wang, Y.; Zang, Y. An improved approach to detection of amplitude of low-frequency fluctuation (ALFF) for resting-state fMRI: Fractional ALFF. *J. Neurosci. Methods* **2008**, *172*, 137–141. [CrossRef]
43. Heni, M.; Kullmann, S.; Ketterer, C.; Guthoff, M.; Linder, K.; Wagner, R.; Stingl, K.; Veit, R.; Staiger, H.; Häring, H.U.; et al. Nasal insulin changes peripheral insulin sensitivity simultaneously with altered activity in homeostatic and reward-related human brain regions. *Diabetologia* **2012**, *55*, 1773–1782. [CrossRef] [PubMed]
44. Tang, Y.Y.; Tang, R.; Posner, M.I. Brief meditation training induces smoking reduction. *Proc. Natl. Acad. Sci. USA* **2013**, *110*, 13971–13975. [CrossRef]
45. Chen, A.C.; Oathes, D.J.; Chang, C.; Bradley, T.; Zhou, Z.W.; Williams, L.M.; Glover, G.H.; Deisseroth, K.; Etkin, A. Causal interactions between fronto-parietal central executive and default-mode networks in humans. *Proc. Natl. Acad. Sci. USA* **2013**, *110*, 19944–19949. [CrossRef]
46. Wang, G.Z.; Belgard, T.G.; Mao, D.; Chen, L.; Berto, S.; Preuss, T.M.; Lu, H.; Geschwind, D.H.; Konopka, G. Correspondence between resting-state activity and brain gene expression. *Neuron* **2015**, *88*, 659–666. [CrossRef]
47. Kong, F.; Hu, S.; Wang, X.; Song, Y.; Liu, J. Neural correlates of the happy life: The amplitude of spontaneous low frequency fluctuations predicts subjective well-being. *Neuroimage* **2015**, *107*, 136–145. [CrossRef] [PubMed]
48. Holmes, A.J.; Hollinshead, M.O.; O'keefe, T.M.; Petrov, V.I.; Fariello, G.R.; Wald, L.L.; Fischl, B.; Rosen, B.R.; Mair, R.W.; Roffman, J.L.; et al. Brain Genomics Superstruct Project initial data release with structural, functional, and behavioral measures. *Sci. Data* **2015**, *2*, 1–16. [CrossRef]
49. Turner, J.A.; Damaraju, E.; Van ERP, T.G.; Mathalon, D.H.; Ford, J.M.; Voyvodic, J.; Mueller, B.A.; Belger, A.; Bustillo, J.; McEwen, S.C.; et al. A multi-site resting state fMRI study on the amplitude of low frequency fluctuations in schizophrenia. *Front. Neurosci.* **2013**, *7*, 137. [CrossRef]
50. Comon, P. Independent component analysis, a new concept? *Signal Process.* **1994**, *36*, 287–314. [CrossRef]
51. Beckmann, C.F.; DeLuca, M.; Devlin, J.T.; Smith, S.M. Investigations into resting-state connectivity using independent component analysis. *Philos. Trans. R. Soc. B Biol. Sci.* **2005**, *360*, 1001–1013. [CrossRef] [PubMed]
52. De Pasquale, F.; Della Penna, S.; Snyder, A.Z.; Lewis, C.; Mantini, D.; Marzetti, L.; Belardinelli, P.; Ciancetta, L.; Pizzella, V.; Romani, G.L.; et al. Temporal dynamics of spontaneous MEG activity in brain networks. *Proc. Natl. Acad. Sci. USA* **2010**, *107*, 6040–6045. [CrossRef] [PubMed]
53. Calhoun, V.D.; Adali, T. Multisubject independent component analysis of fMRI: A decade of intrinsic networks, default mode, and neurodiagnostic discovery. *IEEE Rev. Biomed. Eng.* **2012**, *5*, 60–73. [CrossRef]
54. Li, Y.O.; Yang, F.G.; Nguyen, C.T.; Cooper, S.R.; LaHue, S.C.; Venugopal, S.; Mukherjee, P. Independent component analysis of DTI reveals multivariate microstructural correlations of white matter in the human brain. *Hum. Brain Mapp.* **2012**, *33*, 1431–1451. [CrossRef] [PubMed]
55. Long, Q.; Bhinge, S.; Calhoun, V.D.; Adali, T. Independent vector analysis for common subspace analysis: Application to multi-subject fMRI data yields meaningful subgroups of schizophrenia. *NeuroImage* **2020**, *216*, 116872. [CrossRef] [PubMed]
56. McKeown, M.J.; Makeig, S.; Brown, G.G.; Jung, T.P.; Kindermann, S.S.; Bell, A.J.; Sejnowski, T.J. Analysis of fMRI data by blind separation into independent spatial components. *Hum. Brain Mapp.* **1998**, *6*, 160–188. [CrossRef]
57. Beckmann, C.F. Modelling with independent components. *NeuroImage* **2012**, *62*, 891–901. [CrossRef] [PubMed]
58. Calhoun, V.D.; Adali, T. Group ICA of fMRI Toolbox (GIFT). 2004. Available online: https://trendscenter.org/software/ (accessed on 23 February 2019).
59. Abou-Elseoud, A.; Starck, T.; Remes, J.; Nikkinen, J.; Tervonen, O.; Kiviniemi, V. The effect of model order selection in group PICA. *Hum. Brain Mapp.* **2010**, *31*, 1207–1216. [CrossRef] [PubMed]

60. Correa, N.M.; Adali, T.; Li, Y.; Calhoun, V.D. Canonical correlation analysis for data fusion and group inferences. *IEEE Signal Process. Mag.* **2010**, *27*, 39–50. [CrossRef]
61. Chen, J.; Calhoun, V.D.; Liu, J. ICA order selection based on consistency: Application to genotype data. In Proceedings of the 2012 Annual International Conference of the IEEE Engineering in Medicine and Biology Society, San Diego, CA, USA, 28 August–1 September 2012; pp. 360–363.
62. Laney, J.; Adali, T.; Waller, S.M.; Westlake, K.P. Quantifying motor recovery after stroke using independent vector analysis and graph-theoretical analysis. *NeuroImage Clin.* **2015**, *8*, 298–304. [CrossRef]
63. Bell, A.J.; Sejnowski, T.J. An information-maximization approach to blind separation and blind deconvolution. *Neural Comput.* **1995**, *7*, 1129–1159. [CrossRef]
64. Correa, N.M.; Adali, T.; Li, Y.; Calhoun, V.D. Comparison of blind source separation algorithms for fMRI using a new Matlab toolbox: GIFT. In Proceedings of the (ICASSP '05), IEEE International Conference on Acoustics, Speech, and Signal Processing, Philadelphia, PA, USA, 23 March 2005; Volume 5, p. v-401.
65. Long, Q.; Jia, C.; Boukouvalas, Z.; Gabrielson, B.; Emge, D.; Adali, T. Consistent run selection for independent component analysis: Application to fMRI analysis. In Proceedings of the 2018 IEEE International Conference on Acoustics, Speech and Signal Processing (ICASSP), Calgary, AB, Canada, 15–20 April 2018; pp. 2581–2585.
66. Mori, S.; Wakana, S.; Van Zijl, P.C.; Nagae-Poetscher, L. *MRI Atlas of Human White Matter*; Elsevier: Amsterdam, The Netherlands, 2005.
67. Wakana, S.; Caprihan, A.; Panzenboeck, M.M.; Fallon, J.H.; Perry, M.; Gollub, R.L.; Hua, K.; Zhang, J.; Jiang, H.; Dubey, P.; et al. Reproducibility of quantitative tractography methods applied to cerebral white matter. *NeuroImage* **2007**, *36*, 630–644. [CrossRef]
68. Hua, K.; Zhang, J.; Wakana, S.; Jiang, H.; Li, X.; Reich, D.S.; Calabresi, P.A.; Pekar, J.J.; van Zijl, P.C.; Mori, S. Tract probability maps in stereotaxic spaces: Analyses of white matter anatomy and tract-specific quantification. *NeuroImage* **2008**, *39*, 336–347. [CrossRef]
69. Kelly, R.E., Jr.; Alexopoulos, G.S.; Wang, Z.; Gunning, F.M.; Murphy, C.F.; Morimoto, S.S.; Kanellopoulos, D.; Jia, Z.; Lim, K.O.; Hoptman, M.J. Visual inspection of independent components: Defining a procedure for artifact removal from fMRI data. *J. Neurosci. Methods* **2010**, *189*, 233–245. [CrossRef] [PubMed]
70. Calhoun, V.D.; Adali, T. Unmixing fMRI with independent component analysis. *IEEE Eng. Med. Biol. Mag.* **2006**, *25*, 79–90. [CrossRef]
71. Ray, K.L.; McKay, D.R.; Fox, P.M.; Riedel, M.C.; Uecker, A.M.; Beckmann, C.F.; Smith, S.M.; Fox, P.T.; Laird, A. ICA model order selection of task co-activation networks. *Front. Neurosci.* **2013**, *7*, 237. [CrossRef]
72. Feis, R.A.; Smith, S.M.; Filippini, N.; Douaud, G.; Dopper, E.G.; Heise, V.; Trachtenberg, A.J.; van Swieten, J.C.; van Buchem, M.A.; Rombouts, S.A.; et al. ICA-based artifact removal diminishes scan site differences in multi-center resting-state fMRI. *Front. Neurosci.* **2015**, *9*, 395. [CrossRef]
73. Tohka, J.; Foerde, K.; Aron, A.R.; Tom, S.M.; Toga, A.W.; Poldrack, R.A. Automatic independent component labeling for artifact removal in fMRI. *NeuroImage* **2008**, *39*, 1227–1245. [CrossRef] [PubMed]
74. Bhaganagarapu, K.; Jackson, G.D.; Abbott, D.F. An automated method for identifying artifact in independent component analysis of resting-state fMRI. *Front. Hum. Neurosci.* **2013**, *7*, 343. [CrossRef]
75. Griffanti, L.; Douaud, G.; Bijsterbosch, J.; Evangelisti, S.; Alfaro-Almagro, F.; Glasser, M.F.; Duff, E.P.; Fitzgibbon, S.; Westphal, R.; Carone, D.; et al. Hand classification of fMRI ICA noise components. *NeuroImage* **2017**, *154*, 188–205. [CrossRef]
76. Power, J.D.; Plitt, M.; Laumann, T.O.; Martin, A. Sources and implications of whole-brain fMRI signals in humans. *NeuroImage* **2017**, *146*, 609–625. [CrossRef] [PubMed]
77. Kassinopoulos, M.; Mitsis, G.D. Identification of physiological response functions to correct for fluctuations in resting-state fMRI related to heart rate and respiration. *NeuroImage* **2019**, *202*, 116150. [CrossRef] [PubMed]
78. Liu, J.; Pearlson, G.; Windemuth, A.; Ruano, G.; Perrone-Bizzozero, N.I.; Calhoun, V.D. Combining fMRI and SNP data to investigate connections between brain function and genetics using parallel ICA. *Hum. Brain Mapp.* **2009**, *30*, 241–255. [CrossRef] [PubMed]
79. Bressler, S.L.; Menon, V. Large-scale brain networks in cognition: Emerging methods and principles. *Trends Cogn. Sci.* **2010**, *14*, 277–290. [CrossRef]
80. Pessoa, L. Understanding brain networks and brain organization. *Phys. Life Rev.* **2014**, *11*, 400–435. [CrossRef]
81. Hwang, K.; Bertolero, M.A.; Liu, W.B.; D'esposito, M. The human thalamus is an integrative hub for functional brain networks. *J. Neurosci.* **2017**, *37*, 5594–5607. [CrossRef] [PubMed]
82. Young, K.A.; Manaye, K.F.; Liang, C.L.; Hicks, P.B.; German, D.C. Reduced number of mediodorsal and anterior thalamic neurons in schizophrenia. *Biol. Psychiatry* **2000**, *47*, 944–953. [CrossRef]
83. Sherman, S.M.; Guillery, R.W. *Exploring the Thalamus and Its Role in Cortical Function*; MIT Press: Cambridge, MA, USA, 2006.
84. McIntosh, A.M.; Maniega, S.M.; Lymer, G.K.S.; McKirdy, J.; Hall, J.; Sussmann, J.E.; Bastin, M.E.; Clayden, J.D.; Johnstone, E.C.; Lawrie, S.M. White matter tractography in bipolar disorder and schizophrenia. *Biol. Psychiatry* **2008**, *64*, 1088–1092. [CrossRef] [PubMed]
85. Mamah, D.; Conturo, T.E.; Harms, M.P.; Akbudak, E.; Wang, L.; McMichael, A.R.; Gado, M.H.; Barch, D.M.; Csernansky, J.G. Anterior thalamic radiation integrity in schizophrenia: A diffusion-tensor imaging study. *Psychiatry Res. Neuroimaging* **2010**, *183*, 144–150. [CrossRef] [PubMed]

86. Ćurčić-Blake, B.; Nanetti, L.; van der Meer, L.; Cerliani, L.; Renken, R.; Pijnenborg, G.H.; Aleman, A. Not on speaking terms: Hallucinations and structural network disconnectivity in schizophrenia. *Brain Struct. Funct.* **2015**, *220*, 407–418. [CrossRef] [PubMed]
87. Koshiyama, D.; Fukunaga, M.; Okada, N.; Morita, K.; Nemoto, K.; Yamashita, F.; Yamamori, H.; Yasuda, Y.; Fujimoto, M.; Kelly, S.; et al. Role of frontal white matter and corpus callosum on social function in schizophrenia. *Schizophr. Res.* **2018**, *202*, 180–187. [CrossRef] [PubMed]
88. Epstein, K.A.; Cullen, K.R.; Mueller, B.A.; Robinson, P.; Lee, S.; Kumra, S. White matter abnormalities and cognitive impairment in early-onset schizophrenia-spectrum disorders. *J. Am. Acad. Child Adolesc. Psychiatry* **2014**, *53*, 362–372. [CrossRef]
89. Karlsgodt, K.H.; van ERP, T.G.; Poldrack, R.A.; Bearden, C.E.; Nuechterlein, K.H.; Cannon, T.D. Diffusion tensor imaging of the superior longitudinal fasciculus and working memory in recent-onset schizophrenia. *Biol. Psychiatry* **2008**, *63*, 512–518. [CrossRef]
90. Glahn, D.C.; Laird, A.R.; Ellison-Wright, I.; Thelen, S.M.; Robinson, J.L.; Lancaster, J.L.; Bullmore, E.; Fox, P.T. Meta-analysis of gray matter anomalies in schizophrenia: Application of anatomic likelihood estimation and network analysis. *Biol. Psychiatry* **2008**, *64*, 774–781. [CrossRef]
91. Zhou, S.Y.; Suzuki, M.; Hagino, H.; Takahashi, T.; Kawasaki, Y.; Matsui, M.; Seto, H.; Kurachi, M. Volumetric analysis of sulci/gyri-defined in vivo frontal lobe regions in schizophrenia: Precentral gyrus, cingulate gyrus, and prefrontal region. *Psychiatry Res. Neuroimaging* **2005**, *139*, 127–139. [CrossRef]
92. Barta, P.E.; Pearlson, G.D.; Powers, R.E.; Richards, S.S.; Tune, L.E. Auditory hallucinations and smaller superior temporal gyral volume in schizophrenia. *Am. J. Psychiatry* **1990**, *147*, 1457–1462.
93. Job, D.E.; Whalley, H.C.; McConnell, S.; Glabus, M.; Johnstone, E.C.; Lawrie, S.M. Structural gray matter differences between first-episode schizophrenics and normal controls using voxel-based morphometry. *NeuroImage* **2002**, *17*, 880–889. [CrossRef] [PubMed]
94. Onitsuka, T.; Shenton, M.E.; Salisbury, D.F.; Dickey, C.C.; Kasai, K.; Toner, S.K.; Frumin, M.; Kikinis, R.; Jolesz, F.A.; McCarley, R.W. Middle and inferior temporal gyrus gray matter volume abnormalities in chronic schizophrenia: An MRI study. *Am. J. Psychiatry* **2004**, *161*, 1603–1611. [CrossRef]
95. Kuroki, N.; Shenton, M.E.; Salisbury, D.F.; Hirayasu, Y.; Onitsuka, T.; Ersner, H.; Yurgelun-Todd, D.; Kikinis, R.; Jolesz, F.A.; McCarley, R.W. Middle and inferior temporal gyrus gray matter volume abnormalities in first-episode schizophrenia: An MRI study. *Am. J. Psychiatry* **2006**, *163*, 2103–2110. [CrossRef]
96. Cascella, N.G.; Fieldstone, S.C.; Rao, V.A.; Pearlson, G.D.; Sawa, A.; Schretlen, D.J. Gray-matter abnormalities in deficit schizophrenia. *Schizophr. Res.* **2010**, *120*, 63–70. [CrossRef] [PubMed]
97. Onitsuka, T.; Shenton, M.E.; Kasai, K.; Nestor, P.G.; Toner, S.K.; Kikinis, R.; Jolesz, F.A.; McCarley, R.W. Fusiform gyrus volume reduction and facial recognition in chronic schizophrenia. *Arch. Gen. Psychiatry* **2003**, *60*, 349–355. [CrossRef] [PubMed]

MDPI
St. Alban-Anlage 66
4052 Basel
Switzerland
www.mdpi.com

Applied Sciences Editorial Office
E-mail: applsci@mdpi.com
www.mdpi.com/journal/applsci

www.ingramcontent.com/pod-product-compliance
Lightning Source LLC
Chambersburg PA
CBHW061556210926
51850CB00009B/2120
* 9 7 8 3 0 3 6 5 9 6 3 8 9 *